# Blueprint

How Our Childhoods Make
Us Who We Are

## Lucy Maddox

ROBINSON

ROBINSON

First published in Great Britain in 2018 by Robinson

This edition published in 2020 by Robinson

1 3 5 7 9 10 8 6 4 2

A CIP catalogue record for this book
is available from the British Library.

ISBN: 978-1-47213-789-0

Typeset in Adobe Jenson Pro by SX Composing DTP, Rayleigh, Essex
Printed and bound in Great Britain by Clays Ltd, Elcograf S.p.A.

Papers used by Robinson are from well-managed forests
and other responsible sources.

MIX
Paper from
responsible sources
FSC® C104740

Robinson
An imprint of
Little, Brown Book Group
Carmelite House
50 Victoria Embankment
London EC4Y 0DZ

An Hachette UK Company
www.hachette.co.uk

www.littlebrown.co.uk

# Contents

# Introduction

When I taught at the Anna Freud Centre it was in the library of a tall house at number 21 Maresfield Gardens, London. It had a polished wooden floor, shelves full of books and psychoanalytic journals, a quietly brilliant librarian, and pictures and statuettes of Freud dotted about. The first time I arrived to lecture, a bust of Freud watched me from the mantelpiece as I got the computer and projector set up. He did not look friendly. He was not good for my feelings of imposter syndrome.

I decided to befriend Freud, though, and at the start of each successive academic year that I returned to the library he felt more welcoming. He no longer seemed to glare at me. In my head I had a bit more of a laugh with him.

I lectured there for five years, alongside my clinical job in a psychiatric ward for teenagers at South London and Maudsley NHS Trust. I taught how to consider child development and childhood difficulties from multiple theoretical perspectives. Some Freudian ideas were in there, but also many other ways of thinking, including the study of neuroscience, cognition, behaviour, family systems – a rich and diverse range of approaches. These were approaches I'd been lucky enough to be taught myself, not only through my own studies but also through working with talented and knowledgeable colleagues and supervisors in research or clinical roles. As I researched and taught (and listened to visiting speakers, several of whom are interviewed in this book), I couldn't help but relate every topic I lectured on or heard about to myself and who I was. To try and make sense of how I had become who I am, and to consider what I could learn from the studies of how we all grow and develop, in childhood and beyond – it was irresistible. Throughout it all I was continually surprised at how few of my non-psychology friends knew about the really classic studies, the juicy morsels from developmental psychology, that shed so much light on who we are as adults.

I think everyone should know about this stuff. Just as we are taught about the famous experiments of Newton and Archimedes, shouldn't we also be told how the vital importance of affection was discovered, and how we develop our sense of self? Psychology isn't taught in school in the same way that Physics or Chemistry is, so unless you've chosen to do A-level Psychology or read a book related to the subject then there's no reason you would ever come across the classic experiments. Some psychology studies have captured the public imagination – the Milgram study in which people gave each other electric shocks, for example, or the Stanford Prison experiment where people were asked to behave as guards and prisoners. Yet few of these notorious examples are relevant to what goes on when we are children, and how we develop into who we are as adults.

Every single one of us has been a child. The roots of our adult selves go right back to our first experiences. How we think, act and interact is influenced by our early years, yet most people don't know the key findings from child development. By thinking about these we can learn to notice and understand how we tend to be in relationships, in times of stress or change, or when faced with tricky decisions.

What do we do with the knowledge? We don't have to do anything. Just by knowing it we might be kinder and more accepting of ourselves, and of other people. We might start to clock some of the patterns we fall into as a result of how our early years shaped us. And maybe try out some different choices. Or not. Either is OK.

How do we get a sense of identity? How are our romantic styles affected by our early relationships? How do we develop an ability to think in complex ways, manage moral dilemmas and motivate ourselves and others?

This book tries to explain some of the key ideas about how we become ourselves. It includes recent studies that are still ongoing and some of the really classic experiments. From babyhood to adulthood, it aims to weave together cutting-edge research, everyday experience and clinical examples.

Draw a blob of lipstick on the nose of a child who is younger than about eighteen months, and show them their reflection in a mirror, and they tend to reach out to touch the reflected red dot, or even search behind the mirror for the other child. From about eighteen months, children reach to touch their own nose instead, understanding that the image in the mirror shows a reflection – the beginnings of a sense of identity. How does this rudimentary sense of self develop into the ability to have existential crises at three in the morning as we try to work out who we are and what we are doing? And does our identity really get fixed in adolescence, or do we have the potential to change?

Baby monkeys separated at birth from their real mothers will prefer to be with a cloth-covered soft dummy mother than with a harsh wire monkey that they can't cuddle, even when the wire mother is the one with the milk bottle. Before the classic study that established this paradox, affection and closeness were thought to be unimportant for parenting. This study and others revealed that love is as much a primary need as food. The types of bonds we develop with our caregivers as children are key to our ability to develop as adults, and they influence the way we approach romantic relationships later on. Understanding this can radically change the way we understand ourselves and our interactions with others.

I've divided this book into chapters based on different areas of development, so it isn't chronological, but organised by topic. Each chapter goes through some of the most significant experiments in that field and shows how they contribute to our understanding of who we are as adults. For many of the topics I've interviewed some of the experts to get their points of view, and I've included more recent studies too, to try to give you an overview of where the current understanding has reached. If you want to skip ahead to a different chapter you can: each one can be read as a standalone piece in its own right.

This book is far from a parenting guide or child psychology textbook: it is written for anyone who is interested in how our childhood can affect our adulthood. Having said that, parents or those working professionally with children might find it doubly interesting – to think both about their own development and about that of the children they are raising or working with.

Ultimately, to understand a bit more about what goes on in the landscape of childhood gives us a better chance of understanding who we are now. This book aims to be a road map to the bits of your development you probably don't even know about. Bon voyage.

# 1

# Before we are even born

The developing foetus floats about in amniotic fluid like a tiny astronaut. There in the dark, no smells or tastes, it is increasingly sensitive to sounds, tactile sensations and its mother's emotions. It's hard to imagine what it must be like for it as it grows into an entity with awareness. What starts as a blob of cells transforms into a mini-person, all happening there in the water-cushioned darkness.

Our baby brains are built while we are in the womb. What begins as a 3-millimetre-long neural tube develops into a brain with 10 billion brain cells (or neurons) and 100 trillion connections. Every minute that we are in the womb 250,000 neurons are being created. Neurogenesis, or the birth of brain cells, continues long after our own birth. The connections between brain cells also continue to be formed and pruned for years after we emerge.

As the foetus floats, brain cells are created, move through the foetal brain to the area where they need to be, and once in situ start to change into the type of neuron needed in that specific brain region. They begin to connect with other cells around them, until they are part of a massive web of brain cells. This phenomenal process happens so quickly that it resembles the spread of a rampant plant like bindweed

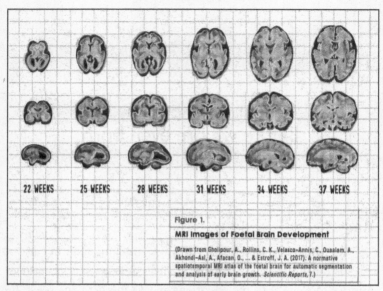

Figure 1: MRI Images of Foetal Brain Development, from Gholipour, A., Rollins, C. K., Velasco-Annis, C., Ouaalam, A., Akhondi-Asl, A., Afacan, O., ... and Estroff, J. A. (2017). A normative spatiotemporal MRI atlas of the foetal brain for automatic segmentation and analysis of early brain growth. *Scientific Reports*, 7.

or Japanese knotweed. Between twenty and forty weeks old our brain grows as much in size as it does from when we are born to when we are five years old. Magnetic resonance images between twenty-four and forty weeks show the brain starting off as a smooth shape, and developing the characteristic walnut-esque folds and wrinkles as it develops and changes.

## Getting conscious

It is what these added brain-wrinkles confer that is important. Development in the foetal brain is directly related to one of the biggest mysteries of humanity: consciousness. Just how much conscious awareness a foetus has is a contentious issue which impacts on debates around abortion.

Professor Vivette Glover has spent most of her career studying the foetus, and in particular the interaction between foetal and maternal experiences. Interviewed in her home in London, colourful paintings on the walls around her, she explains how she became interested in foetal development, and what her studies can teach us about some of the mysteries of this tiny bundle of cells. A biologist by training, Glover had been working on postnatal depression at Queen Charlotte's Maternity Hospital when a Professor of Foetal Medicine asked her to switch tack and help him study whether foetuses experience pain.

'We found we could measure foetal stress responses,' says Glover. 'Some of the [medical] procedures involved putting a needle through the foetal tummy, and we could measure stress hormones in the blood as this was happening. We showed that the foetus from about 18 weeks did show stress responses.'

'I became interested in the experience of the foetus,' Glover goes on, 'and that led me to start to think about long-term effects. Does what happens in the womb have long term effects on foetal neurodevelopment?'

Foetuses respond to touch at ten weeks, but this is a reflex response: at this stage they don't have enough of a nervous system to feel anything. 'At the beginning it's just a blob, and it barely has a nervous system,' says Glover. 'If you have a 12-week foetus and you touch its foot it will react, but there isn't any connection between the periphery and the brain, so we think it's just a reflex like a knee-jerk. I think it's extremely unlikely it's feeling anything.'

It is hard to pinpoint a single point where a foetus 'wakes up' into conscious awareness. As we grow in the womb we develop an ability to feel and sense in different domains, and this ability is very dependent on how many weeks post-conception we are, or our gestational age.

'We know that the older foetus is the same age as a pre-term baby,' says Glover. 'If you watch pre-term babies they're very reactive. They're reactive to being held by their mothers, to sounds, and so on. And it seems highly likely that the older foetus, the last trimester, is experiencing quite a lot. Now there's an interest in when might they start to become conscious, start to become aware . . .' Glover explains that the foetus's peripheral nervous system, including its spinal cord, doesn't start to link up with its brain until about seventeen weeks. 'It's not fully linked up till about 26 weeks, so in that sort of period, between 17 and 26 weeks, I think the foetus is probably starting to have experiences. By the last trimester then it's likely to have a whole range of experiences.'

By twenty-six weeks old foetuses can respond to sound, and it is quite probable that by this point they have a degree of consciousness – perhaps from even earlier, at about twenty weeks. At around this twenty- to twenty-six-week mark they might be able to feel pain. Later on in their development foetuses can certainly hear, and they seem to be able to retain a memory of some sounds for a few days after birth. There's something quite weird and amazing about being able to imagine the developing foetus becoming increasingly aware in there. A gradual waking-up.

One elegant experiment, albeit with small numbers, compared the babies of seven women who regularly watched the Australian soap opera *Neighbours* with eight women who didn't.[1] Two- to four-day-old babies who had heard the theme tune while in the womb became alert, stopped moving about and had a reduced heart rate when they were played the tune again. They didn't react in this way to other unfamiliar pieces of music, and they hadn't been exposed to the *Neighbours* theme tune after birth, so they must have recognised it from hearing it *in utero*. Babies of the mums who didn't watch *Neighbours* didn't react when the theme tune was played to them. It's not that there's anything particular about the *Neighbours* tune itself, but that the foetus has learned to associate it, presumably with a sense of pleasure in its mother, relaxing in front of a favourite show.

Another systematic study[2] took a range of different frequencies of sound, and played them, using a speaker, to 450 foetuses of different ages through their mothers' abdomens, while monitoring the foetus using real-time ultrasound equipment. This study found that the first age at which a foetus responded to a sound was nineteen weeks, but this was only a single foetus from the whole sample population. As gestational age (the number of weeks since conception) increased, the more foetuses were able to hear the sounds, with 96 per cent of foetuses responding (by moving) to the lower-frequency sounds by the age of twenty-seven weeks.

I love the image this conjures up – as though the foetuses are dancing along to the sounds. The higher-frequency sounds weren't responded to until the foetuses were older, with the first responses to 1,000Hz and 3,000Hz being at twenty-nine and thirty-one weeks respectively. All of the foetuses moved in response to the 1,000Hz sounds when they were thirty-three weeks, and to the 3,000Hz at thirty-five weeks. As the foetuses matured they needed the sounds to be played less loudly, suggesting their hearing was improving.[3]

## Nutrition, drugs and alcohol

Whether or not we've been exposed to the *Neighbours* theme tune as a foetus is unlikely to cause us any harm, but some of what we experience in the womb can have longer-lasting and more negative effects.

Several major diseases which tend to occur later on in life, including coronary heart disease, hypertension and type 2 diabetes, have been linked to impaired growth and development during critical periods of time in the womb. The idea that diseases later on can be caused by things that happen at critical, sensitive periods in early life has been called 'programming', and researchers think this happens because of permanent consequences to structure, physiology and metabolism which result from normal development of the bodily systems being interrupted.

One longitudinal study of 25,000 men and women in the UK has looked at the size people are at birth, and compared this with diseases that occur in middle age. People who were born small overall, or particularly short or thin, had higher rates of coronary heart diseases, higher blood pressure, higher cholesterol and abnormal glucose-insulin metabolism. Findings were linked only to size at birth, not to whether people were born prematurely, and researchers suggest that it is the lack of nutrients at critical developmental periods that causes the problem.[4]

It is a bit like baking a cake and opening the oven door too early when the cake is still rising: if we interrupt the process of foetal development by disrupting the nutrients available then we can affect the outcome much later on. As we develop in the womb, there are critical periods for different areas of development, where our foetal self is especially sensitive to the *in-uterine* environment. If the womb environment changes in these periods, specific outcomes might be affected and our development might head off down a different trajectory. It's for this reason that clinicians who are seeing a child or adolescent, or sometimes even an adult, might ask a lot about how things were for the person before birth, even though it might seem like an odd set of questions. Some difficulties which become apparent later on might have very early origins.

It's not just not eating enough nutrients that can have a negative effect. The extreme consequences of mothers ingesting some drugs have been widely documented. One example comes from the devastating effects of thalidomide, prescribed to pregnant women in the late 1950s and early 1960s to prevent nausea, which resulted in severe disabilities from birth in thousands of children. Thalidomide had been tested in rats and found to be safe, but hadn't been tested in other species, and the huge effects on foetal development that the drug had in humans led to drug-testing procedures being changed to include at least two species, one of which isn't a rat.[5]

Recreational drugs can also affect foetal development, although researching these effects can be difficult, as there are often lots of confounding factors. A mother who is using one recreational drug is more likely to be using others, including nicotine and alcohol. She may also be more subject to the effects of poverty, stressful living circumstances and possibly a long-term history of difficult or traumatic circumstances. These external stressors might have been what initially prompted illegal drug use, and alongside the effects of addiction might be what makes it hard to give up drugs despite being pregnant.

In the 1990s there was a lot of media coverage of 'crack babies', the term given to babies whose mothers were using crack cocaine during the pregnancy. Worries about the ability of these children to concentrate, or to regulate their emotions and control their behaviour, were widespread, although at least one larger-scale review[6] did not find robust evidence for this.

Effects of maternal opiate consumption on the baby – for example if the woman is using heroin while pregnant – are more clearly seen. These babies are smaller on average, and when they are born have to go through the physical symptoms of withdrawal from heroin. Conclusions in relation to cannabis use during pregnancy are less clear, as too for amphetamine use.

One of the most popular recreational drugs is, of course, alcohol. Whether or not to have a glass of wine in pregnancy is a hotly debated topic, but the effects of alcohol abuse in pregnancy really do relate to alcohol *abuse*, not to an occasional small amount. Foetal alcohol syndrome (FAS) has wide-ranging and potentially devastating consequences, including babies being born smaller, having dysfunctional effects on their central nervous system development and recognisable facial abnormalities. The exact level of alcohol exposure that leads to this is unknown, although some studies place it at about 70g of absolute alcohol daily.[7]

## How are you feeling, Mum?

While the dangers of mothers' behaviour during pregnancy are widely spoken about, and many people have opinions about drinking or drug-taking during the pregnancy, a less talked-about influence is stress: the effect of how the mother is feeling while she's pregnant.

As early as the nineteenth century, the poet Samuel Taylor Coleridge pretty much nailed the importance of *in utero* experience. Years before the neuroscience was around to back him up he was writing: 'Yes! The history of man for the nine months preceding his birth would probably be far more interesting and contain events of greater moment, than all the three score and ten years that follow it.'[8] Contemporary National Child Welfare campaigns in the early 1900s emphasised the importance of mothers staying calm: 'Worry, fear and anger may affect his mother's blood, which supplies his food. Therefore she should be calm, happy and sweet-tempered.'[9] Obviously easier said than done when you are carrying a whole other life form inside of you.

There are so many pieces of advice for pregnant women, some of which are conflicting: which cheese to eat, whether a glass of wine is OK, whether Guinness is good or not, how to speak to the bump . . . It turns out the most important thing is probably counter to what these reams of advice produce: as stress-free an environment as possible for the mum. Our mother's emotional state during pregnancy can contribute to how we are as children, and even later on how we are as adults.

At the same time as Professor Vivette Glover was beginning to study foetal experience, a growing body of animal research – which is still prolific today – was showing that stressing a pregnant animal has long-term effects on the development and behaviour of its offspring.[10] There was very little research then into the same possibility in humans. Decades later, the research has moved on a long way.[11]

'There's no doubt that if the mother is anxious or depressed or stressed whilst she's pregnant, this increases the risk for the child having a whole lot of different problems,' says Vivette Glover. She doesn't want women to worry unnecessarily, however: 'There's nothing inevitable about it, and actually most children aren't affected . . . But it definitely increases the risk of a range of different problems.'

Risks associated with increased maternal stress include a greater chance that the child will be anxious or depressed, will be diagnosed with ADHD or conduct disorder, or will have learning difficulties. It is also associated with a greater likelihood of pre-term delivery, and being born early and a bit smaller comes with its own range of challenges.

Just because something confers an increased risk doesn't mean we know how big a risk it is. Glover's group has done research with the large population study based in Bristol, UK, called Children of the 90s or ALSPAC, which stands for Avon Longitudinal Study of Parents and Children. The study has followed 14,500 families for the last twenty-five years, providing a massive amount of data on child development, and Glover has used the data to study how big an effect it is on a child if their mum is sad or anxious during pregnancy.

'Because it's such a large study we can allow for all the other things that may be affecting child development,' explains Glover. 'If the mother is anxious or depressed while she's pregnant she's likely to be anxious or depressed postnatally and it could affect her parenting, and we know that there are certainly post-natal effects on the child. But using ALSPAC we're able to allow for that and try and identify the component of what happens in the womb.'

Glover's research found that if the mother is in the top 15 per cent for having symptoms of anxiety or depression herself, this doubles the risk of her child having a probable mental disorder at age thirteen. 'It doubles the risk from about 6 per cent to about 12 per cent.'

Even if 12 per cent of children are affected, then approximately 90 per cent of children still don't go on to have serious problems, despite their mums being stressed during pregnancy, so it is by no means inevitable that there will be a detrimental effect. But a doubling of risk is still significant, and anything we could do to reduce that increased risk could help a large number of children and mothers feel better.

The study with the mothers from ALSPAC found different effects on child mental health at thirteen depending on how much stress the mother had reported. They initially looked just at the outcome of the children of the mums who scored in the top 15 per cent for anxiety and depression – the ones who were feeling really bad – but when the researchers looked across the whole range of scores for maternal anxiety and depression, they found a 'dose response curve'. The mother even being a little bit anxious or depressed increased the risk of a mental-health problem in the children at age thirteen.

How stressed is stressed, though? Pregnancy is inherently stressful. Suddenly being responsible for a tiny person growing inside you, then worrying about the pregnancy, birth and what happens next, coping with bodily changes and the different capability you now have to do the things you used to – all of this means most mums-to-be will be a bit stressed. Glover recognises this and doesn't want to make things even worse by adding to the list of concerns. Worrying about the effects of worry just creates even more stress.

## What helps?

'There is always a problem in talking about this,' says Glover. 'We don't want to worry women more. The whole thing is we want to get more support, give them more support.' There's a lot we can do, she thinks, when we know someone who is pregnant, and a lot in particular that partners can do: 'Fathers are probably the most important person in all this. It's been shown that fathers giving social support is extremely helpful in buffering against the mother feeling symptoms of anxiety or depression, so a supportive partner is really helpful – talking to them, being there, helping, just generally feeling that they are there in support, willing to listen.' Crucially, the reverse is also actively unhelpful: 'An abusive or unhelpful partner adds to stress. We have research showing that if the mother said the partner was emotionally cruel to her, this increased the risk of the child having emotional and cognitive problems later. So the partner is the most important person either for good or for bad.' Glover thinks that professionals caring for pregnant women ought to do a lot more to bring partners in. 'The father can feel quite isolated at this time when the mother gets pregnant. He can feel he's going to get left out, what's his role, so it can be a difficult time for fathers. We ought to do much more to involve fathers, and help them understand how important they are, even during the period of pregnancy.'

Fathers are important, but they aren't the only ones. Depending on family circumstances, same-sex partners and other sources of support can also be crucial. 'Grandparents have a big role to play,' says Glover, who is herself a grandmother. 'Society in general needs to be more supportive, more understanding of this. There isn't really much research on it, but it seems likely that all that will help. And employers ought to be understanding too. A lot of women want to carry on working, but they want to work in the way they want to work, so that they feel in control.'

While the research on the effects of stress during pregnancy is growing, there has so far been less appetite for looking at the opposite end of the spectrum: whether there are positive effects from mothers experiencing calm or happiness during pregnancy. Studies on interventions suggest that if depression is reduced by cognitive behavioural therapy during pregnancy then outcomes for the child improve,[12] but there is nothing on the effects of women being particularly happy or tranquil when pregnant.

## How does the foetus sense the mother's stress?

During all of this the foetus is floating about in its dark bubble. It might feel intuitive that the foetus would be affected by how its mother feels, but when we stop to think about it, how does the foetus feel the difference? How can it witness its mother's emotions?

'We're only just beginning to understand this,' says Vivette Glover. 'In animals there has been lots more research, but we're not rats and our biology is different, so it all has to be established in humans.'

This all gets quite intricate in its biology, but it's worth hanging on in there. What has been looked at most is the system that makes the stress hormone cortisol. 'In animals this is clearly shown to be involved,' explains Glover. 'In humans it is less clear, because when we're pregnant the placenta puts out a lot of cortisol anyway, which changes the function of the mother. We are starting to look at lot at the placenta. The placenta filters what chemicals are coming from the mother and go through to the foetus, and it's very sensitive to the signals it gets from the mother.' Glover and others are demonstrating that if the mother is more anxious or depressed, the level of the enzyme in the placenta that breaks down the stress hormone cortisol goes down, potentially allowing more stress hormone to travel from mother to foetus.

'The placenta is clearly quite a major part of what's going on,' says Glover, 'but we are very far from fully understanding this. Does it matter if the foetal brain is exposed to more cortisol? Well, we're starting to have some studies where we recruited women having amniocentesis and we measured the level of cortisol in the amniotic fluid, and we brought the babies back when they were eighteen months and tested the baby's IQ. We found that the higher the level of cortisol in the womb, in the amniotic fluid, the worse they did in the baby IQ test.'

## Baby IQ Tests

Baby IQ tests make it sound like babies are sat down with paper and pen and asked to do some maths problems. In fact, these tests are designed especially for infants to try to assess how well they are understanding the world around them. One of these is called the Bayley's,[13] where a toddler is given a range of puzzles, including being shown pictures and asked to point at specific animals, or asked to post shapes through similar-shaped holes. This mixture of understanding words and physical manipulations is a rough guide to the level at which a baby is thinking.

In contrast to Vivette Glover's work on maternal stress and mental health outcomes in children, some studies have shown that a small amount of maternal stress seems to be associated with slightly better scores in some aspects of baby IQ tests.[14] 'It depends a lot what you're looking at,' acknowledges Glover, 'and maybe for some outcomes a bit of stress is actually helpful . . . For other outcomes, emotional behavioural ones, I don't think any stress is helpful.'

## Attachment and stress

Glover and her colleagues also used a measure of the relationship between the mothers and babies in the study on the effects of cortisol. They used something called 'the strange situation' (described in detail in Chapter 2), which attempts to measure how mother and baby relate to each other, and in particular how the baby copes with separations and reunions.

'We found that if they were securely attached' (see Chapter 2), 'this completely buffered the effects of being exposed to more cortisol in the womb, whereas if they weren't, there was quite a strong relationship. It's showing what we already knew, but in this particular case very strongly: that it's not all over at birth: that what happens postnatally is just as important . . . Sensitive mothering and a securely attached infant can counteract some of the effects of what happens in the womb.'

Glover is now getting to see again the children she recruited early on in their lives, post-amniocentesis, and is using the brain-scanning technique of functional magnetic resonance imaging, which looks at brain activity during different tasks. She is looking to see if there are differences in brain function between babies who had more cortisol in the womb, and those who had less. Initial studies suggest that there are differences, with increased cortisol early on seeming to produce similar functional changes to those in the brains of children with a diagnosis of attention deficit hyperactivity disorder (ADHD), a diagnosis associated with high levels of activity and a difficulty concentrating.

## Contagious stress and the mysteries of the placenta

While Glover's field of research is going through a time of rapid growth and discovery, there remain a number of unsolved mysteries, especially about the biology of what exactly goes on to transmit stress from mother to foetus.

Scientists are beginning to understand some of the changes in the placenta that happen when mums are stressed, but less is known about what changes in the mother to cause these changes in the placenta. The main suspect is a group of chemicals called the pro-inflammatory cytokines, known to be important in pregnancy in general, and starting to be implicated in changes to mothers' moods.

Another mystery is how quickly foetuses can sense how stressed their mothers are. One study, by Catherine Monk at Columbia University, involved mothers in the last trimester of their pregnancy (with foetuses who were between six and nine months old). Monk asked the pregnant women to do some tests: some mental arithmetic problems (counting backwards from 1,000 in seventeens), and one classic psychology test called the Stroop task.

The Stroop task presents conflicting information. Colour words (e.g. red, blue, yellow) are written in a different colour to the one they name. So, for example, the word blue is written in red ink. Participants are asked to either name the colour the words are written in, or read the word aloud. Usually people are slower when the information is conflicting, and quicker at responding when the information is consistent, for example if the word blue is written in blue ink.

This task, and the mental arithmetic problems, can be slightly stressful. Monk monitored foetal heart rate while the women were performing these tasks, and then asked the mothers afterwards whether they had been stressed out. About half the women said they had found the tasks stressful, and half said they hadn't been too bothered and it had felt more like a game. When mothers said it was stressful, the foetal heart rate tended to increase while the mums were doing the tasks and then come down afterwards. In the other foetuses there wasn't much change.

The mechanism for this happening is unknown. 'Cortisol, which we're particularly interested in', explains Vivette Glover, 'takes about twenty minutes to rise' – yet the heart rate increases straight away. Could it be a rise in noradrenaline, a common culprit in stress reactions? No: 'Noradrenaline gets totally broken down by the placenta,' says Glover, 'so it's not noradrenaline. Maybe it's a physical tenseness – we just don't know. Something's going on, but we don't know what.' So far this is another unsolved mystery.

## What would be the use of catching stress off our mums?

Mothers who are stressed in pregnancy are more likely to have children who are fearful. The most reproduced finding is that these children are more anxious. The second most reproduced finding is an increase in the diagnosis of ADHD.

Glover and colleagues think they understand why: 'If you think in evolutionary terms, then when mothers who were our ancestors were feeling stressed, it was probably because there was a real, physical threat out there: there were more dangerous animals around, or more dangerous tribes, and so on. And if you're more vigilant and more anxious, that actually can help you detect the threat quickly, and could even be protective. The same with ADHD: having a readily distracted attention means that you hear the rustle in the undergrowth, you notice something's happening. It's helpful to be readily distracted. Whereas in our modern society, when children are at school and are supposed to be learning, being distracted by something outside the window is really maladaptive. But in evolutionary terms both the vigilance with the anxiety, and the readily distracted attention, might be protective.'

This makes sense of how stress caused in this way might be contagious, but it is less helpful at understanding why this doesn't happen to all children of stressed mothers. Lots of children don't have raised anxiety or symptoms of ADHD, even though their mothers were very stressed in their pregnancies.

## Nature and nurture

The way children are cared for postnatally is one factor in how they go on to develop, and can have a huge impact. Another factor is one we can do less about: our genetic vulnerability to develop various difficulties and react to stress in certain ways.

Vivette Glover takes us through some of the detail. In the case of ADHD, there's a gene that codes for an enzyme that breaks down lots of different neurotransmitters, including adrenaline, noradrenaline and dopamine. We all have different forms of that gene, and if we have one particular form and our mother was anxious in pregnancy then we are much more likely to develop ADHD as a child. If we have one of the other two forms we seem to be resistant to it. There's not just one gene involved usually – for example, schizophrenia is associated with over 150 different genes – so the effects are complicated, but the interplay of nature and nurture can be seen quite clearly.

Taking this to the next level is the study of epigenetics (see Chapter 4): the study of changes in gene expression which get passed down through the generations. If our grandmother was stressed when she was pregnant with our mum, does this have an effect on us? The animal research suggests it does, but the human research is less established.

If it does turn out to be the case that effects can be passed on through generations, then it emphasises even more the need for good care of pregnant women.

## Why are some women more stressed in pregnancy?

Reasons for mothers being more stressed in pregnancy often relate to their circumstances at the time. Women in an abusive relationship are bound to be highly stressed. Mothers experiencing financial hardship or the isolation of being without a supportive partner or family are also likely to be stressed. Some of these difficulties also make it harder to meet the unborn child's nutritional needs.

Some reasons for stress might relate to circumstances that go further back into the past. 'We're starting to think about early exposure to trauma,' says Vivette Glover, 'and there's some evidence that early abuse is quite a big factor in this. If the woman's been exposed to any sexual abuse, for example, there's double the risk of her being depressed or anxious while she's pregnant.' Women might remember sexual trauma more when they are having to be physically examined, or might be more likely to think about their own childhood because they are about to become a mother themselves.

The potential long-term effects of sexual trauma in childhood are really thrown into relief by this sort of study. It seems so tragically unfair that someone who has experienced sexual abuse might, through no fault of their own, then be more likely to be stressed by pregnancy and potentially have a child with more difficulties. This vicious cycle isn't inevitable – after all, there are lots of babies who are resilient to maternal stress *in utero* – but it does suggest that we should be doing more as a society to ensure that all mothers have adequate care and support. To give everyone the best possible chance of developing as healthily as possible we need to be helping mums-to-be to feel as good as possible, and giving extra support if needed.

## Looking to the future

For Vivette Glover, the most important area of future research is looking to see which interventions can help women who are anxious, depressed or stressed in pregnancy. 'There doesn't always have to be a direct match between their feelings and their biology,' she says. 'We know anyway that in pregnancy it's the biology that must be changing, that must affect the placenta, that must affect the foetus.' Whether this means psychological interventions, practical interventions or medication-related interventions, getting in there early could have an important effect. Screening mums-to-be and new mums for any signs of low mood or anxiety is crucial, and questionnaires like the Edinburgh Postnatal Depression Scale[15] are routinely used to check for any problems, and to help women spot symptoms of depression that they might be inadvertently minimising. Despite the routine use of such screening, Glover thinks that depression in pregnancy is massively under-detected, and that when it is detected it is mostly not treated. 'There's huge room for improvement in both detection and then treatment,' she says. 'Depression is the major complication of pregnancy.' Glover is interested in large-scale interventions that can help large groups of women, and especially internationally applicable interventions such as the use of music.

Ultimately, if we're interested in trying to make sure the future generations have as good a start as possible, the best thing we can be doing is be aware of the stresses of pregnancy, and do all we can to support pregnant women and new mums. Both practical and emotional support are important. Women need to have adequate resources to be able to feed their unborn child and themselves, and to live free from fear and violence. Women also need to be able to talk about how they feel without this causing them more guilt or worry, and without anyone making them feel unusual or abnormal if they suffer from anxiety or depression at this time. Fathers (and mothers' partners if they are in a same-sex relationship) also need our support in the run-up to and after birth. There are many people involved in raising a child, and the greater the web of support around them, the better things are likely to go.

# 2
# Getting attached

The Beatles sang that love is all you need. And although it's not quite this simple there's a grain of truth in it. The established pre-war view of what we need as children growing up included the practical things: the basics of food, shelter and protection from harm. Very little attention was given to love and affection, which were seen as the cherries on the cake, a nice extra but certainly not the main ingredient.

We now think virtually the opposite. Having a loving relationship with a parent or other caregiver as we grow up is one of the most important things – for our well-being when we are children and for how we function as adults. These first relationships, we now recognise, affect our ability to regulate our emotions, interact with others, and even the way we behave in romantic situations. How did the established wisdom get turned on its head?

## Running geese

Some of what we know about the effect of parenting comes from watching animals. An ethologist called Karl Lorenz spent a long time in the 1930s watching greylag geese. He hatched and cared for them himself, feeding them and looking after them, away from their mother. The geese 'imprinted' on Lorenz, following him in a line wherever he went, running when he ran, swimming when he swam.[1] This imprinting behaviour seemed to occur during a critical period after hatching, causing some researchers to wonder whether a similar bond is formed in infant children.

## Wire monkeys

At around the same time, at Stanford University, Harry Harlow was observing monkeys. In a series of experiments, which would never get past an ethics committee today, Harlow separated baby rhesus monkeys from their mothers, and raised them in separate cages. He allowed the baby monkeys access to two models of a larger monkey: one made only of wire, but with a bottle of milk attached, and one with no milk attached, but which was covered in a soft, terry-toweling-type material. The motherless monkeys spent all their time on the soft mother, craving the softness, and only went to the wire mother for food, before quickly returning to the towelled surrogate.[2] This called into question all previous ideas about food and shelter being the primary drives for an infant, and suggested that the role of comfort might be much more important than was previously thought.

## Getting attached

We often talk about 'getting attached' to someone or something, but the psychological definition of attachment is quite specific, and refers to the 'deep and enduring emotional bond that connects one individual to another through time and space',[3,4] and in particular in relation to the bond between a caregiver and infant. Most babies and their caregivers form an attachment, and the quality of this attachment can be affected by the sort of care the baby experiences. For infants, attachment manifests as an innate primary drive to stay close to their caregiver. We now know that these early attachment relationships form the basis, to some degree, for the way we relate to others as adults.

The father of the idea of attachment was John Bowlby, a psychiatrist, psychologist and psychoanalyst, who was interested in what happened to children who were separated from their caregivers early on. Bowlby himself was raised mostly by nannies, seeing his mother for an hour a day, as was the norm in upper-class English families in the early 1900s.[5] Bowlby went to boarding school at the age of nine, and these early experiences might have made him especially interested in the effects of early maternal relationships, although he never spoke publicly about this, except to say he had been 'sufficiently hurt but not sufficiently damaged' by these formative years.

Bowlby's multiple trainings meant he was influenced by neurobiology, ethology and psychodynamic theory, and it was his work which was responsible for a shift from thinking of infant–mother attachment as being a secondary drive to it being a more important, primary drive. He was especially influenced by the study of what happened when attachment was interrupted. He worked with several young delinquents, his 'forty-four thieves', as he called them, who were forty-four young people with 'delinquent behaviour' from Canonbury in London. Bowlby noticed an association between theft in young people and loss of a caregiver early on, which led him to think about how early experiences of loss can have profound later effects.

Bowlby outlined his ideas on attachment in three volumes called *On Attachment and Loss*. His ideas were radical at the time. At their heart was his emphasis on the basic drive of a child to be near its caregiver. In 1958 he outlined how he thought instinctual behaviours from the infant elicit a caregiving response from their parent or carer, and in 1969 he elaborated this into a description of a goal-corrected system triggered by environmental cues, where, if an infant feels unsafe in its surroundings, it cries or seeks comfort in a way which brings the caregiver straight back to it.

Attachment, Bowlby recognised, needs some cognitive understanding of object permanence – that is, objects or people that go away can also come back again, and haven't simply disappeared – and also some capability for recognition of the parent. Because of this need for some cognitive capacity, Bowlby thought that infants between the ages of nought and two months were in a pre-attachment stage, infants between two and seven months had the beginnings of recognition, and infants of about seven months and older began to get anxious about strangers and want their main caregiver. This explains why very young babies can often be happily passed around, but at about the age of seven or eight months they get really upset if you hand them to someone they don't know. By the age of two years old the attachment relationship has evolved into the goal-corrected partnership Bowlby described between infant and caregiver, with both caregiver and infant driven to be close to the other, particularly if something frightening happens in the environment around them. The infant develops increased independence as it grows to have an internal working model of this attachment relationship, which means having a mental representation of its caregiver as a permanently existing person who comes back even if they go away.

## Robertson and the two-year-old hospital patient

John Bowlby's ideas were picked up by James Robertson, who applied them to children's experiences of hospital in the UK in the mid-twentieth century, and whose work in turn fed into Bowlby's understanding. Back then, visiting hours in hospitals were really restricted, so parents could only come and see their children for a couple of hours a day on certain days, or in some hospitals not at all for the first month.

Robertson got increasingly interested in studying the effects of what happened when children were suddenly separated from their parents like this. Robertson visited and observed, describing and filming the immense distress that young children showed when they first went into hospital, followed by a period of withdrawal and 'settling down', during which they became compliant with whatever doctors and nurses wanted them to do. Far from seeing this compliance as a good thing, Robertson saw it as a sign of danger and institutionalisation. He characterised a child's reaction to separation in hospital as following three phases: protest, despair and detachment.

Robertson started studying separations of children and caregivers in the late 1940s and through the 1950s, when it was common for sick children to be admitted to hospital for as long as it took for them to be well again. My mum recalls her brother being taken into hospital on his own as a child for polio, with treatment that included being in an iron lung. Nowadays that kind of solitary hospital experience for children just wouldn't happen. Parents are allowed to stay in hospital with their young children: there is much more recognition of how important that affectional tie is.

## Strange situation

John Bowlby's work on the importance of attachment and loss of caregiving figures influenced his colleague Mary Ainsworth, who worked with him in the 1950s. In the 1960s she went off to Uganda to develop a way of measuring the quality of attachment between a caregiver and child, which is still used today. The 'strange situation', as it's called, certainly is strange, and involves observing a child's reaction to their caregiver leaving and arriving in a room, and also their reaction to a stranger.

Created in 1963, the 'strange situation' involves seven short episodes. Ainsworth repeated it with American children in Baltimore, so it's also a measure which, unusually, was developed in both Africa and North America, as opposed to solely being Western. The whole thing takes about twenty minutes. There is a standardised structure, so it matters which order you do it in. The aim is to be able to observe separations and reunions between children and their caregivers, and reactions to a stranger. The structure goes like this:

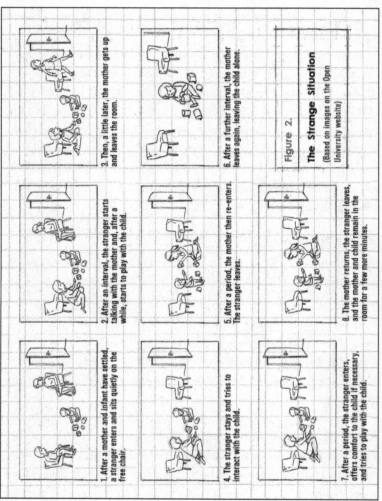

Figure 2: The Strange Situation (based on images on the Open University website)

1. Mother and child sit in a room together for three minutes.

2. A stranger enters, sits for one minute, talks to mum for one minute, and plays with the child for one minute.

3. Mother leaves.

4. The stranger tries to play with the child.

5. Mother returns, the stranger leaves, mother settles child and stays for three minutes.

6. Mother leaves the child alone for up to three minutes.

7. Stranger enters, tries to settle the child, withdraws if possible. This lasts three minutes.

8. Mother returns, stranger leaves, mother settles the child and sits down.

Ainsworth painstakingly observed infants and caregivers in this odd paradigm. Strange though it is, it did provide a systematic way of observing the attachment relationship that Bowlby had described. Her research highlighted that attachment isn't just there or not there. She observed different qualities of attachment, and her work vastly increased the range and subtlety of the concept.

The way in which different children responded to the 'strange situation' could, thought Ainsworth, be reliably categorised into three main classifications; later on, a fourth classification got added.

### Secure

'Securely attached' children get upset when their caregiver leaves the room, and are comforted when they return. That sounds so obvious, but it's not what all infants do.

## Insecure ambivalent (sometimes called insecure-resistant)

These children are observed to be preoccupied or clingy with the caregiver before they leave the room. When the caregiver returns, the child can't leave them easily, and finds it hard to settle down to play.

## Insecure-avoidant

'Insecure-avoidant' children seem not to notice when their caregiver leaves the room, but if you monitor their physiology using heart-rate monitors or other measures of stress like the stress chemical cortisol, their physical signs show that they are stressed.[6]

## Disorganised

There were also a number of children who didn't group so easily into a pattern, and some years later a researcher called Mary Main got interested in these. She grouped them into a category that reflected the way they didn't respond reliably and called this 'disorganised'.[7] This is the only attachment classification that is really worrying – it is seen in a high percentage of children who have experienced child maltreatment.

Infants in the 'disorganised' category tend to display unusual reactions to separation and reunion with a parent or caregiver. They often get distressed when the adult leaves the room, but then they stay distressed when the parent comes back. Their responses to their caregiver are incoherent. They might approach the caregiver, then move past them, or they might sometimes freeze, crying, unsure of whether to reach for the adult or stay away. They might bang their head or engage in other behaviours that seem bizarre or upsetting.

In the case of children who have experienced maltreatment from that adult, they are faced with the dilemma of wanting to seek comfort, but knowing that the adult is also possibly the source of danger. If the adults who are looking after you are also the ones responsible for hurting you, then you are in an impossible position, where feeling safe is unfamiliar, where literally anything could happen. During my clinical career I've seen disorganised attachment reactions in young children with their parents less than a handful of times, but it's very distressing and disturbing to witness.

Mary Main did a lot of work to create interview templates: the Child Attachment Interview (CAI) and the Adult Attachment Interview (AAI) enable children and adults to be classified as having experienced one of the different attachment patterns, since the 'strange situation' is only for infants and caregivers. Main's focus was on the idea that our early attachment relationships become internalised to give us working models for other, later relationships, including friendships and romances.

## Fortune-telling from attachment styles

What does it mean to be each of the different attachment styles? Studies vary in how many people they estimate to be in each classification, but not all of us are securely attached, and it doesn't mean we're in real trouble if we're not. Although there are bigger variations within cultures than between them, there is also some evidence that certain sorts of attachment style are found more often in certain cultures,[8] probably because of cultural variation in child-rearing practices.

It clearly sounds a lot better to be classified as having a secure relationship with a caregiver than an insecure one, but in fact the only really worrying category is disorganised, which predicts psychopathology at the age of seventeen,[9] and predicts internalising problems (like anxiety and depression) and externalising problems (like behavioural disturbance) later on to a fairly reliable degree. Other insecure attachment classifications seem slightly predictive of later problems, but only to a very small degree. Similarly, some studies have found positive associations between secure attachment and better abilities to form close friendships with other school-age children, compared to insecure, but the associations are small. Different attachment classifications will result in different ways of interacting socially and with teachers at school, and some evidence suggests securely attached infants might find the transition to school easier, but this is far from cut and dried (see Chapter 14 for more).

Feminist critiques of the ideas of attachment have at times suggested that the original theories place too much emphasis on mothers' relationships with their children, with the political agenda of meaning that mothers have to stay at home. In fact, cross-cultural studies looking at different family set-ups show that it doesn't have to be the mother who has the strong affectional tie with the baby. Although it often is the mother, it can also be the father and other family members – in the case of children raised in Israeli kibbutzim, for example, it can be multiple caregivers.

## How long do these classifications last?

If we're classified at the age of two as having a certain type of attachment relationship, does it mean we'll have similar ideas about relationships later on? There is some evidence that attachment classifications affect the relationships that teenagers have with their parents. Michael Dykas[10] and his team interviewed 189 seventeen-year-olds and their parents. They found that secure adolescents recollected memories of conflict with their parents more favourably over time than insecurely classified adolescents, and that mothers of secure teenagers also remembered things more positively.

This difference in how conflict was remembered could be an example of the internal working model of the parent–child relationship affecting the way memories were viewed in retrospect, i.e. that having a better general idea of how our relationship is with our parent or child makes us remember even tricky times more favourably. It could also be that more secure attachments early on predict easier parental relationships later.

Another longitudinal study found that 70 per cent of attachment classifications remained the same twenty years later, and that changes in classification were related to big life events, often negative, like abuse, parental divorce or major loss.[11] Despite this, other theorists, Main and Crittenden[12] among them, suggest that although our experiences are viewed through the lens of our early attachments and subsequent models of relationships, we can also use new experiences to adapt our internal working models. This is, in a way, the basis of psychotherapy.

Professor Peter Fonagy is a psychoanalyst and clinical psychologist who works in many different roles in the UK, including chief executive of the Anna Freud Centre and joint head of the Clinical Psychology training course at University College London. I interviewed him about his views on attachment, and whether attachment classifications could change over time.

'I would very strongly disagree with any kind of scenario that suggests irreversibility,' he began. 'I think attachment is far more context-dependent than we used to think. The reality is that secure attachment is more a state than it is something that characterizes an individual. So a mother in a particular social context will be there for that infant, and will be profoundly bound to that infant, but if, for example, she anticipates, in countries like Ethiopia and Brazil, that that infant is likely to die of malnutrition, she actually disengages from the infant. So I would now say that social environment is likely to trigger different attachment styles which are adaptive.'

Fonagy views this hopefully. 'That of course means that there is a tremendous amount of flexibility in the system. I'm not saying that early experience doesn't bias one towards particular styles of relationship. What I'm saying is that actually sometimes extreme environments can switch these tendencies off or on.'

For Fonagy, this makes the potential role of therapeutic work even more important. He thinks work with parents can help foster strong attachment relationships – 'by creating, for example, a much more supportive environment for a caregiver who is struggling to develop a bond with their child, I think we can turn off the kind of emergency mode that she's functioning in that makes her rather hesitant to develop attachment relationships. And by us clearing the debris, as it were, in the way of that biological link, she or he's able to actually form an attentive, sensitive, caring relationship, which eluded that person in the past.'

Fonagy doesn't see attachment classifications as something it is helpful to use as a permanent label. 'Environment plays a tremendously important role in triggering a number of attachment styles that are available probably to all of us,' he explains. 'We have a range of templates for managing close relationships. Some are dominated by trust, by mutuality, by a capacity to give comfort and to receive it; others by wanting to maintain a bigger distance; others again by trying to amplify one's need for reassurance and support – perhaps by exaggerating one's signals of distress. All these are styles we have that can be triggered by specific environmental contexts, and I think we mustn't underestimate the importance of that social context – the behaviour of other people – in terms of driving or inhibiting one or more of these patterns.'

Some of this Fonagy sees as a potential outcome of good-quality therapy, although he thinks the remit of therapy goes further than this. 'The change that occurs in therapy, I think, is through the establishment of a trusting relationship in an individual who is struggling with that. Perhaps because of early experiences of neglect or trauma they are untrusting of their environment, and therefore they don't learn from their social environment, and maintain perhaps maladaptive views of expectations about others. They appear rigid or unchanging, and people get frustrated with them. They go into therapy, they have a relationship with a person that they can trust, because they feel understood and recognised and paid attention to, and that gradually makes them shift their attitude towards their wider social world. How therapy works is perhaps not so much by changing the person in the consulting room, but by enabling the social world around them to be perceived as more benign, as more friendly, and that changes the individual.'

## Intimacy and romance

Romantic love has long been a fascination of the arts: poetry, painting, songs, plays, films – you name it and love has been the subject of it. If it were as simple as a magazine quiz then it would all have been mapped out long ago. But it isn't. Science has shied away from the subject for a long time, but psychology and more recently neuroscience have now begun to tackle it.

There does seem to be some relationship between how we were attached with our caregivers and how we are in romantic relationships. The networks in the body and brain which are involved in early attachment are also activated when we fall in love, and some of the same brain chemicals are involved. This notion of the attachment system being 're-activated' means we might be more likely to fall into familiar patterns of relating when we feel strongly about someone.[13] And those patterns aren't always helpful.

There isn't just one type of love – a fair few flavours have been identified. In *Colours of Love* J. A. Lee lists three 'primary love styles': *eros* (romantic, passionate love), *ludus* (game-playing love), and *storge* (friendship love); and three secondary styles: *mania* (possessive, dependent love), *pragma* (logical, 'shopping-list' love), and *agape* (selfless, all-giving love).[14] It has been suggested that if we ignore *pragma* and *storge* as not being to do with romance, then attachment classifications could be linked to the others – secure attachment with *eros* and *agape*, avoidant attachment to *ludus* and ambivalent attachment to *mania*.[15]

Donatella Marraziti and Stefano Baroni, two Italian neuroscientists, reviewed the literature of love in relation to the brain, to try and sort out what we know and what we don't know.[16] Their description of romantic love is not the most poetic: they pragmatically call it 'a form of bonding which is crucial for the survival of the human species'. The idea of picking a partner to create a new family with, they point out, means we have to seek out an individual outside of our original family group. 'This is a real paradox,' they point out: 'humans are attracted to, courted by and breed with genetically not-related individuals whom they would otherwise instinctively avoid.' Romantic love, then, is a psychological strategy which enables us to overcome the instinctive fear of a new other person, and instead lets us create a deep, often lasting bond with a stranger, with the result of producing healthier offspring.

Put like this, romantic love seems like an awful trick of nature – getting us to abandon all safety and hitch our wagon to a strange horse, all in the name of procreation. According to the authors, the big and uncontrollable feelings, the weird behaviours we get into, and the uniquely human capacity to be aware of the whole caboodle, this is all just to do with our brain chemistry getting us to find a partner we can propagate our genes with. This is what love is. Less romantic, yes, than the way the arts have explored love. But maybe useful in a different way.

The arts do much to describe what falling in love *feels* like. Have a think about it. Can you recall a time when you were falling love? When I think about falling in love I recall feeling addicted to a person, wanting to see them, to touch them, to be near them. The arts do less explaining about *why* we might fall in love.

In contrast, neuroscience does less to describe love but may be a helpful way of understanding the neural basis for this feeling. The way that neuroscience labels love, when it does go there, echoes romantic poems about the madness of love, but in a much more diagnostic tone. Marraziti and Baroni compare love's initial characteristics as akin to several diagnoses of mental-health problems. 'Some years ago we demonstrated that Obsessive Compulsive Disorder and Romantic Love share a common dysfunction,' they write. They talk first of all about the period they call 'attraction', which they say lasts between six months and three years. They compare the highs and lows of attraction to the mood swings of bipolar disorder, and they compare the intense craving of a partner to compulsions, and an over-valuing of the potential love object as the most wonderful being in the world. They draw the comparison between OCD and love, considering the intrusive thoughts about the person we are so enamoured of – unable to stop our thoughts returning to them, obsessively. They also describe that first phase of attraction as a stressful time and, conversely, as something which can come out of stressful situations.

## Neurotransmitter snap

The chemical culprits for romantic attraction are the same brain chemicals involved in the disorders that the Italians refer to. Dopamine, serotonin, cortisol and noradrenaline are all implicated in those mental illnesses, and are all also implicated in falling in love. The authors distinguish between the phase of initial attraction and the subsequent phase after the initial madness of falling in love passes, when a more stable and comforting relationship develops. They relate this more to the idea of attachment, and call romantic attachment the 'glue' that keeps a relationship together. This phase has been more associated with oxytocin and vasopressin, which by no coincidence are brain chemicals that are also released when mothers (and fathers) have skin-to-skin contact with their babies.[17, 18] Injecting oxytocin into the brains of rats who have been given rat pups to foster brings on maternal behaviours,[19] and the idea of oxytocin as a 'prosocial molecule' has captured researchers' imaginations worldwide and led to some fascinating experiments.

You may not think of the vole as being a romantic animal, but prairie voles have been widely studied for their monogamous tendencies. Giving prairie voles oxytocin, after they had been left in a room with another prairie vole for six hours (without being allowed to mate), kick-starts monogamous bonding between the two. Separate studies of voles who have mated, compared with voles who haven't, showed increased receptors in the brain for oxytocin and vasopressin in those voles who have already mated. These increased levels of oxytocin and vasopressin were associated with monogamy, and an increased tendency for the male partner to defend the female.[20, 21]

It is deliciously intuitive that similar molecules might be involved in our first important relationships: with our main caregivers, and with our important relationships later on. It is perhaps slightly too tempting to believe in a 'love molecule', and Patricia Churchland and Piotr Winkelman urge some caution.[22]

They point out that it is more likely for a brain chemical to be having a generalised effect across large systems than to be having a specific effect on sophisticated behaviours related to romance. They think it is more likely that oxytocin increases partner bonding by increasing a feeling of security and decreasing anxiety. They think the evidence on how oxytocin is transmitted across the blood brain barrier, from the body into the brain, suggests it can't be having the quick effects that some scientists claim. In other words, as is often the case in affairs of the heart, it's complicated.

Nevertheless, whether the neurotransmitters are having a broad-brush effect or more of a specific one, there is evidence that oxytocin and vasopressin are implicated in the same systems that are involved when we are attached to early caregivers as when we are in love later on. It makes sense, then, both from a brain-chemistry perspective and from a theoretical perspective, that we might fall into ways of relating to partners which are influenced by the relationships we had with our parents. It may very well not be a conscious choice, but templates of experience and neurochemistry may influence our reactions to emotional closeness.

## All is not lost

If you think you've had less than optimal attachments as a child, don't throw this book across the room in despair. None of this means that we are doomed. It does mean that it's worth thinking about how we might be primed to respond in relationships, though, and working out if we're happy with the patterns we typically settle into, or if we want to pay some attention to trying to do things differently.

Doom-and-gloom-mongerers who say that everything is determined by our upbringing just aren't doing the field justice. While a lot of what happens to us early on forms part of our blueprint, there is a whole load of genetic predisposition and adult agency that goes along with it to make the particular recipe for who we are. And even if there are bits we don't like but can't really change, knowing what they are and working around them or with them can help a lot.

## Therapy?

Different types of therapy deal with the connection between early relationships and current relationships to different degrees. Psychodynamic psychotherapy in a way goes right for the jugular in this case, working primarily with the relationship that you have in the therapy room with your therapist, and trying to work out how this relates to the relationships you had as a child, and other relationships you have going on as an adult. Trying as well, perhaps, to give you a new model of being, where you feel heard and valued and supported – as though someone is in your corner, but not afraid to point out when you might be repeating patterns.

Cognitive behavioural therapy (CBT) might not be talking as much about the relationship with the therapist, especially if it's a very short-term treatment, but it does take into account how early experiences might set up later thoughts about the world and expectations of others. Systemic therapy draws links, too, just perhaps more explicitly than psychodynamic, and with less about thought patterns than CBT. Ultimately all the therapies (there are many more – see Chapter 19) have some consideration of the relationship we had with our parents, even if it's only a shadow that is acknowledged but not fully unpacked. It certainly doesn't mean it's all our parents' fault, but it does mean that those first relationships are really worth knowing about and thinking about for ourselves, even if that's sometimes painful or difficult.

## 3

# Milestone-spotting
# and going off track

The idea of 'normal' is frighteningly powerful, and both seductive and difficult to deal with. Imagine being told that you are average. Do you feel happy with that? What about being told you are abnormal? Even worse?

Children are terrified of being abnormal. To be labelled 'the weird one' at school is a social disaster, and the same fear can persist well into our adult lives. Wanting to be part of social groups is a primal urge for humans, and from an evolutionary point of view fear of being ostracised makes sense. We need to be part of the pack to survive, and social connectedness is incredibly important for our mental well-being.

But how do we know what's normal or not? As a child, as an adult or as a parent? There is lots of accepted wisdom about what we should be doing at certain ages. While some is a helpful guideline to check that we are developing as we should be, and not needing extra help, other ideas could benefit from being held up to the light and inspected for holes. Our accepted ideas of normal are socially constructed, not absolute.

## What is a child, anyway?

Even our understanding of what constitutes a child is up for grabs. It's hard to pin down what it means that a person is a child. A small person? An un-developed person? A work in progress? Is it purely to do with age, or is there something more about what it is to be a child?

The dictionary defines children in relation to age or stage of physical development: 'A young human being below the age of puberty or below the legal age of majority.'[1] This tells us little about what our expectations of a child are, though, and it turns out that expectations vary widely across decades and across distance.

The Western idea of childhood as a separate stage didn't emerge until the end of the fifteenth century.[2] Before that, children were expected to work, to help in the home, and generally be much like mini-adults. In medieval times the idea of childhood just didn't exist. From then until the nineteenth and twentieth centuries the idea of childhood developed into a more sentimental notion. Expectations that children worked or cared for other children diminished, protection and pleasures became greater, and the landscape of childhood became almost a different world from adult reality.

Other countries and cultures have different ideas: in some parts of Bali, children are called 'small equals', and viewed as divine until 210 days after birth. They are addressed by honorific names, held high, and never placed on the floor until they have undergone a ceremony to acknowledge full entry into the human world.[3] A similarly high status is awarded to children in the Beng of West Africa, where foetuses are viewed as having complete linguistic and social competence, existing in parallel to the spirit world before birth. At birth, babies are thought to have complete spiritual knowledge, which is then lost as they grow up, and are consequently seen to become less competent.[4]

## Children's milestones

The temptation in the West to be able to map out expectations for children is huge. There are apps, wallcharts, books, TV programmes, all dedicated to explaining what children usually do and by when. At best these are interesting and reassuring; at worst they are anxiety-provoking and prescriptive. There are some key milestones that Western children tend to have achieved by certain ages, like speaking, walking and socially interacting. These can be a useful framework for checking out whether there are any developmental difficulties, as long as the context of the child is always taken into consideration.

There are clear age-related expectations for children in the UK and US in terms of typical milestones to be reached in social, physical, cognitive and language development. Parents can get lists or apps that go through these in detail. At one year old, an age none of us will be able to remember for ourselves, our milestones included being shyer with strangers than with our parents, beginning to say simple words like 'Mama' and 'Dada', being able to follow simple instructions like 'Pick up the teddy', and being able to pull ourselves up to stand and walk around using furniture to hold on to (a phenomenon known as 'cruising').

I know from carrying out assessments of children to screen for particular needs or difficulties such as autism spectrum conditions, that taking a detailed developmental history of these kinds of milestones can be very useful. If there is a pattern of several milestones being significantly delayed, alongside current difficulties in social relationships or other areas, it might lead us to wonder whether a child has a developmental disorder, and needs extra input to reach his or her full potential.

On the other hand, when I think about my own milestones, I talked when I was two years old, walked at twelve months and learned to ride a bike in my thirties. This last one especially could be seen as developmental delay, but I don't think I've got a developmental disorder. For every graph that shows a normal track of development, there will be children who develop faster and more slowly, and children who develop more quickly in some areas and more slowly in others.

Some of this depends on what is expected of us, and what our environment allows us to practise.

In a grisly experimental example from the old days of how environment can affect development, David Hubel and Torsten Wiesel used newborn kittens to investigate whether development of the visual system occurs because of environmental stimulation or despite it. They sewed up one eyelid on each of twenty-one newborn kittens, at various different timepoints after birth, and for varying lengths of time. They used two kittens as a normal control, with their eyes untouched. The experimenters found that there was a critical period, within the first ten to twelve weeks. If the kitten's eye was sewn shut during this period, connections between the brain cells in the part of the brain responsible for analysing visual stimuli, the visual cortex, didn't grow as much. There were fewer neurons (brain cells), and fewer connections between those neurons, resulting in impaired vision in that eye. Unsewing the eyelid revealed that the eye itself was fine, but the visual brain areas linked to that eye hadn't fully developed, so the kitten couldn't see out of that eye, even up to five years later. If the kitten wasn't exposed to visual stimuli in that crucial window of time then it was too late.

There are some parallels in the realm of human language development. There is a similar critical period during the first year of life. Initially as babies we can differentiate between speech sounds from all different languages – between phonemes such as 'la' and 'ra' no matter which country we are from. Gradually, as we get older, we find it harder to identify the difference between phonemes that aren't used in our own native language. This leads to difficulties hearing the different sounds in some foreign languages, and certainly difficulties in pronunciation. To truly become fluent in a non-native language we need to be exposed to it as early as possible.

To flip this idea on its head, it's not just about what happens to development in a deprived environment, but also about what happens if you provide extra opportunity for stimulation. Infants who can crawl have depth perception, while those who don't crawl often don't. However, infants who are given opportunities to move about in a stroller develop the ability to perceive depth even if they can't crawl[5] (see Chapter 5). The experience of moving about in the world is important.

Different cultures provide children with different opportunities. Carraher and colleagues interviewed five street children in Brazil who were able to add large sums very accurately in their heads, even though they couldn't do sums on paper.[6] Children of a similar age in the UK struggled to do the same sums. The Brazilian children have had experiences that have enabled them to learn how to do these calculations, and the skill is an important one for their everyday life.

In other parts of the world children may have full-time care of their younger siblings, or be involved in carrying water for long distances or tending cattle for long periods. In the UK this would be considered too much of a burden of responsibility and risk. Different cultural contexts can differ even within the same country, and provide children with different possibilities for learning. A lot of these are class-related, or profession-related, and can be either negative or really positive. A child with parents who are dancers might be exposed to and even taught dance from an earlier age, making it more likely they will be proficient later. Similarly, opportunities to learn can vary wildly across time within the same country as expectations change. In colonial America four-year-old girls knitted stockings and mittens and could embroider intricately. This isn't the case now, although many four-year-olds are proficient at using a tablet computer, a skill that wasn't available before. We are all constantly adapting to a changing world around us, and children perhaps most of all.

## Adult milestones

Milestones are not the preoccupation solely of the anxious parent. We all keep an eye on what others around us are up to, and a check on what we should be doing. In the film of *Bridget Jones's Diary*, Bridget is asked why she thinks people are single. It's probably because we're all covered in scales, she replies. This scene taps right into ideas about what it's normal to be doing at what age, and the idea of 'abnormality' if you're doing something different. The questions we ask of one another reflect timelines we expect people to travel on, and such social expectations often give a weird licence for personal questions we might not otherwise ask. It's common for people to ask newly married couples when they are going to try for children, or to ask long-term partners when they are going to get married. There is an unspoken expectation that we all aspire to go along the same trajectory.

## Family lifecycles

In the late eighties, two family therapists, Carter and McGoldrick,[7] tried to put a framework on what they saw as the normal developmental stages of a family.

Initially they outlined six:

1. Leaving home: the single adult.
2. The joining of families through marriage.
3. Families with young children.
4. Families with adolescents.
5. Launching children and moving on.
6. Families in later life.

Each stage, they reckoned, also has the potential for additional stressful transitions, like the loss of a family member, the addition of a family member (through birth or marriage), re-location of the family, and so on.

It doesn't take much to see that this family life cycle is pretty linear and predictable. It's also very Western, very heteronormative and quite old-fashioned. Carter and McGoldrick were writing about this in the late 1980s, and they fairly quickly realised that it didn't take into account all sorts of things: extended families, elderly parents moving back in, people who don't get married, and of course divorce.

In the wake of this, several different family life cycles have been proposed, including the divorce cycle, the lesbian family life cycle, and all manner of other variations. They all see the process of becoming independent from your family of origin, getting together with someone, and possibly splitting up and getting over this, as being a fairly linear flow.

The problem with this is that there are just so many different ways of living a life. I'm not sure any cyclical or stage-based model ever captures the complexities of our human lives and relationships, and to expect it to do so is at best optimistic, and at worst a bit daft.

More than just not managing to grasp the reality of a situation, these models can end up being actively unhelpful. If our life and our family is doing what the model describes, then great, and it might then be useful to think about common patterns or reactions – but if not, where does that leave us? It leaves us wandering about not only with no clear map of where we are or should go next, but also with a sense that everybody else is following a clear path, and that we have somehow gone off track. Nightmare.

## We're all a bit weird

Just as we have maps of what is supposed to be normal, so we also have manuals of what is supposed to be weird. The International Classification of Diseases (ICD) and the Diagnostic and Statistical Manual (DSM) are bibles of mental illness diagnoses used worldwide as shorthand for clusters of experiences that tend to go together and get in the way of people's lives. Some people like the labels: they find it a relief to open a book that reflects back to them how they have been feeling and behaving, in a neat way that has a name, and is experienced by other people too. The provision of a diagnosis means the hope of a treatment, whether drug or talking therapy. It's a useful shorthand, too, for getting to the nub of someone's difficulties, and for mental-healthcare professionals to be able to quickly summarise to each other broadly what's going on for someone, and what they might need.

On the flip side, diagnosis is a blunt old tool for describing the subtleties of what is happening for someone. It's also highly subjective in several ways. How symptoms are grouped together is not clear-cut, and although the statistical technique of factor analysis is used to see which experiences most commonly cluster together, the results can often be interpreted in lots of different ways. The categories that the DSM lists are debated in meetings of psychiatrists, and the final roll call of 'illnesses' has been decided on as much by clinical opinion as by mathematical rigour. Ultimately, mental illness is harder to pin down in the same way we could describe a physical condition with a clear physical marker. Mental illness is a subjectively experienced condition.

Drawing the line between what is a diagnosable mental illness and what is an extreme emotional human experience is also tricky. Take depression, for example. We all feel sadness, we have probably all known deep despair, a sense of melancholy, or a fleeting and unexplainable sense that nothing is right. Clinical depression is different from feeling sad, but it's a difference of degree. A clinical depression lasts longer, and is so extreme that it stops us functioning – stops us getting out of bed, interacting with people we love, being able to live our lives. There is a spectrum of strength of feeling, and we all lie on it somewhere.

Exactly the same is true for diagnoses that carry even more fear and stigma, like schizophrenia or psychosis. Psychosis is a loss of contact with reality that can involve delusional thinking, hearing or seeing things that other people can't, feeling suspicious of others, or more numbing experiences like feeling less emotional, being able to think less clearly, and lacking motivation to do anything. A long-lasting or recurring psychosis is called schizophrenia, and can be life-crushingly interfering. Some of the experiences associated with psychosis, though, can be had without it developing into schizophrenia. Lots of people hear voices and manage just fine in their lives. Experiences only tip into being considered an illness when they interfere with a person's ability to function, and the difference between someone with a 'mental illness' and someone happening to be having a really terrible time in their lives is sometimes just a matter of who is describing what is happening.

## ADHD or naughty boys?

Attention deficit hyperactivity disorder (ADHD) is a great example of a controversial diagnosis that several professionals argue is socially constructed. Dr Sami Timimi, a consultant child and adolescent psychiatrist, has written extensively about how he thinks the diagnosis is a label that we've given to naughty boys to deal with behaviour which we can't manage. He suggests the diagnosis, and subsequent medication of its symptoms with amphetamines, overlooks the social and cultural factors that contribute to the development of hyperactivity and inattention.

Timimi argues that the social changes we have all undergone over the last century have meant that children's family life, social life, diet and exercise habits are all radically different from before. On top of this, children are increasingly being seen as a consumer market that can be targeted. At the same time, Timimi thinks our tolerance for 'boisterous' behaviour has decreased. The result of all this is, he believes, that it is 'harder than ever to be a normal child or parent'.[8] He sees the rise in medicalised treatment of ADHD as partly an increase in behaviours in children associated with the diagnosis, and partly a change in how society conceptualises ADHD, and indeed what it is to be a child.

Timimi is a brilliant psychiatric renegade, raging eloquently against an ever more pharmaceutically driven diagnosis. It isn't only drug companies who disagree with his thesis, however, and it's much easier to dismiss the diagnosis when you're not living with a child who has been given it. Hyperactivity and difficulty concentrating can be massive problems for some children, which can hugely derail their ability to access education. Drugs are not the only treatment, and indeed parenting interventions are the first-line treatment of choice.[9]

## Diagnostic Spectra

We don't have to dismiss markers of 'abnormality' entirely. More and more, diagnostic labels that were thought of as all-or-nothing are now being thought of on a continuum. Schizophrenia, the loss of contact from reality and associated positive symptoms of delusions and hallucinations, used to seen as something you either had or didn't have. Increasingly nowadays, though, we describe spectra of unusual experiences, and acknowledge that lots of people hear things or see things that others can't hear or see. The importance is the meaning we make of those things, and whether it causes distress or not.

Similarly, with autism spectrum conditions, it is not always clear-cut whether someone has autism or not. It is possible to have social-communication difficulties without reaching the threshold for diagnosis, or for two people to have the same diagnosis of autism spectrum conditions while being at very different degrees of functioning.

In all of this, the important bit is the description of what is actually going on, and whether it's getting in the way of what someone wants to do with their life, or how they are feeling day to day. It's easy for the descriptive, experiential bit to get lost. Perhaps this is partly because the experiences can be hard to describe or explain. It could also be because healthcare and educational support is often diagnostically driven, which makes the labels very powerful as levers for extra support. Such labels also remain powerful in their capacity to stigmatise and shame, despite initiatives to work against this.

## Getting undiagnosed

One interesting thing to consider is how infrequently psychiatric diagnoses in particular get lifted. While it can be hard to be given one, in my clinical experience it is even harder to have one taken away. Diagnosis is an important tool to help people feel understood, and to be clear about the specific problem that is getting in the way of someone's life, but when we look at how some diagnoses are created in the first place, and how often they need their definitions to be revised, I wish we could also be more revisionist in removing them. It is still alarmingly possible for someone to be given a diagnosis of a personality disorder very early on in life, and for this to accompany them throughout the rest of their existence, without a proper pause ever to consider whether it is still relevant or useful.

## Going off track

Whether or not we subscribe to diagnoses, and whether or not we agree with the still-dominant ideas about various stages of life, perhaps the ultimate luxury is the confidence and ability to *choose* to go off track. Developmental milestones in childhood provide a broad-brush picture of which skills we need to be acquiring in order to function. By the time we reach adulthood, all being well, we might be able to choose our own way through the social expectations of our cultures, and decide for ourselves which personal milestones we are aiming for.

## 4

# Epigenetic colouring-in

When I was an undergraduate I studied psychology, philosophy and physiology, a combination that is sadly seldom available. This mix enabled me to think about my interests in both the mind and the body, which probably stemmed partly from my upbringing in the hippy West Country town of Totnes, where crystal healing is abundant and on market days in the high street a bearded man offers gong showers.

As a student, I remember joking with friends about psychology essays, about how hard it was to avoid the formulaic approach of describing the issue, whatever it was, then stating a case for it all being about genetics (nature), then the case for it all being about the environment (nurture), before finally concluding with jazz hands that it was somewhere in between. Of course, the truth is always that it is a mix of different factors, just as now as adults our characters have never been formed by one childhood experience, or by one genetic trait, but instead by a mixture of all the experiences we have had and all the biological tendencies we have inherited.

In the last decade, epigenetics has seemed to colour in some of the big wide space left by these baggy conclusions. Epigenetics is the study of heritable changes in our gene expression which don't involve changes to the underlying DNA sequence itself (for a fuller explanation read on). This relatively new field has offered some tangible mechanisms to explain how nature and nurture might interact, and might continue to interact, down through multiple generations. It is tricky to understand, because it is so much about the biological mechanisms involved, but I think it's worth spending a bit of time with, because it really does seem to be a bit of a missing link in explaining what is going on.

I interviewed researcher in epigenetics Dr Charlotte Cecil at her office in the Henry Wellcome Building at the Institute of Psychiatry, Psychology and Neuroscience. It was in freezing cold January when I met her, and she was in the grip of a horrible cold, but she still managed to explain epigenetics more clearly than anyone I've come across.

## What is epigenetics?

Epigenetics is an umbrella term. It refers to a set of processes that regulate when and where our genes are expressed, which can happen in several different physical ways. People tend to talk about our genes being switched on or off, but for Cecil she's not sure this is the best way of describing what goes on: 'Perhaps if you can think of it sort of as a volume tuner.' Genetic activity is not so much binary, on or off, as something better understood as higher or lower levels of activity, regulated by epigenetic processes. Another metaphor Cecil uses is one from computing: 'Think of your DNA sequence as your biological hardware: you're born with it, it doesn't change, and it provides all of these instructions. Epigenetics can be seen more as the software – the bit that tells your body how to use this hardware.' The instructions that the DNA codes for are for making the proteins that our bodies are made up from: the building blocks to human life, so the way these instructions are 'read' can be vitally important for who we are. Cecil remembers another way of describing epigenetics she read recently, and this is the one that sticks for me: imagine your DNA sequence as a piece of writing: 'epigenetics would be like the punctuation. Which would tell you when to stop, when to pause, and it can change the meaning. So through different cells, you might have different punctuation, so they'd know how to use it [the DNA information] quite differently.' So just as a carefully placed comma can change the meaning of exactly the same words, different epigenetic mechanisms can alter how the same DNA is expressed.

Our genetic make-up stays the same throughout our lifespan. The DNA sequence that we are born with contains all the instructions that are needed for our body to make everything it needs. That genetic code, our individual pattern of the four different letters, the A, T, C and G of the double helix that Crick, Watson and Franklin worked so hard to discover, does not change. But the way that recipe is used, how much it's used and where in the body – that does change over time.

Cecil uses the example of the cells in our bodies. 'Every cell in our body contains exactly the same DNA sequence: the same recipe book for life. And yet we know that cells learn to specialise and perform very different functions. A heart cell, a brain cell or a liver cell – they not only look quite different, but they also perform very different functions. So how is that possible, when they have access to the exact same information? How do they know what bits to read, and what bits they can just safely ignore?' The answer is epigenetics. Epigenetic mechanisms tell cells which bit of genetic material they should read, and which ones they can ignore.

## How does it work?

This can happen through a variety of different mechanisms. The most well-studied has been the process of DNA methylation, mainly because it is a bit easier to measure and observe, and also seems to be more stable. All epigenetic mechanisms work by regulating gene activity:

not by changing anything about the DNA sequence, which stays the same, but by changing how easy or hard it is for that information to be accessed.

## DNA methylation

DNA methylation involves methyl molecules acting like a kind of chemical blanket to wrap around different bits of the genetic material in our cell, to make it harder to read.

If we zoom right into the cells in our body – into the pairs of chromosomes we have in each cell, then further in past the histone structures that our DNA can be wrapped around, we can see the double helix. If we keep zooming in further we come to the base pairs of DNA. Of the four letters which stand for the different molecules involved, A, T, C and G, A always goes with T, and C always goes with G.

Where C and G meet is where methylation can occur.[1] 'DNA methylation is the addition of a methyl molecule to actual DNA base pairs,' explains Cecil, 'but only in the context of the C and G – where C and G meet together.'

'A gene typically has a head called the promoter region,' she goes on, using her forearm and fist as an illustration. 'And it has a body and a sort of tail. What happens is that when the gene is very active, typically you have these things called transcription factors that land on the head of the gene and are able to read all the information in the body and go off and make whatever it is, say testosterone . . . Now, with methylation, what happens is that in the head of this gene you usually have a lot of these CG pairs, called CG islands. This tends to happen around the promotor region, and this is where the methylation can happen. When you have a lot of these methyl molecules that stick to these CG pairs, especially in this promotor region, it creates a sort of chemical blanket, or glue, around the gene.'

This methyl blanket means that even though the information

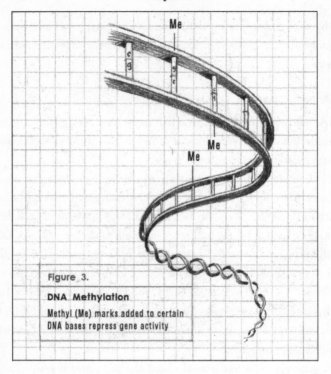

Figure 3. DNA Methylation: Methyl (Me) marks added to certain DNA bases repress gene activity

within the gene hasn't changed, transcription factors can't land any more because of this molecular barrier. Without the methyl groups the gene is open to be accessed, but underneath the blanket it's harder for transcription factors to bind and, because of this, gene expression is reduced. The gene is silenced.

This blanketing isn't a fixed process – it can change over time. 'It's something that basically allows our body to use our genetic information more flexibly,' says Cecil. She is amazed by how much we do with our limited DNA. 'We have less DNA than some fruits, so it's incredible how we actually have quite limited information – 23,000 genes, more or less

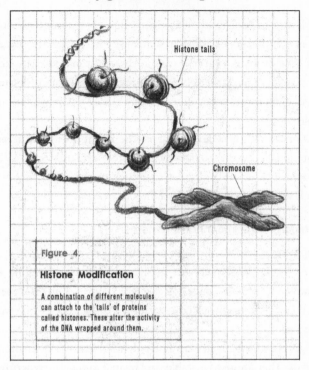

Figure 4. Histone Modification: a combination of different molecules can attach to the 'tails' of proteins called histones. These alter the activity of the DNA wrapped around them

– and we're able to do so much with it. That's because there's a whole additional layer of information that regulates when and how and where these genes are expressed.'

### Histone modification

A less frequently studied mechanism of epigenetics is histone modification. Cecil took me through it from the beginning. 'You have your chromosome that includes your 23 pairs per person, and it contains all your genetic material. If you were to unwind it, you would see that the way it is able to be wound up so tightly into chromosomes is down to what are called histones.' Histones are molecules which are shaped like yo-yos. The DNA is wrapped around these histones like the string of a yo-yo, with a tail of the DNA left sticking out. Molecules can stick to these tails and influence how tightly the DNA is wound around the histone.

'The more tightly wound up it is, the less it is easy to access the information, because it's so tightly compacted,' explains Cecil. 'Whereas you might have other molecules that stick to the DNA and actually loosen it up, thereby opening the DNA to be read.'

## Epigenetics, environment and genetics

The epigenetic mechanisms by which DNA information is made more or less accessible are responsive to what is happening in the environment. So which genes of ours are expressed might depend on our current circumstances, or even our circumstances while we were growing up.

Epigenetic mechanisms are not only affected by the environment around us. They are also heritable, and related to genetic structures. We all have the same genes, but each of us has different versions. Which physical version of the gene we have (e.g. a long version or a short version of the serotonin gene that is linked to mood) can influence how much the gene can be methylated or unmethlyated (blanketed or not) – in other words, how much it is expressed.

Even epigenetic mechanisms, then, are subject to both nature (our genetics) and nurture (our environment). It's pretty mind-blowing.

## Why are people so excited by epigenetics?

Epigenetics is biologically complicated, tricky to understand, but incredibly exciting. It gives us a framework for understanding *how* genes and environment interact.

Charlotte Cecil is very excited by the possibilities. 'If you think of epigenetics as something that tells our body how to develop, and how to use its genetic information, you can see how it's important for so many things. We already knew this was going on, but we thought it was mostly something that just happens internally. What we know now is that, instead, a variety of different environmental influences can actually get under the skin – actually influence the way our genes are regulated.'

This massively adds to the old nature–nurture debate: 'It shows how much this is really a two-way street and a two-way process. Yes, our DNA sequence is there, and can put us at greater or lower risk for certain things, but at the same time the way we live, what we experience, can actually regulate our genetic information.'

Cecil sees this as an adaptive strategy, where our body tries to optimise itself to the needs of its environment. This optimisation can happen for better or for worse: Chapter 9 explains how, when children are exposed to early-life stress, they might adapt in a way that is helpful at the time but problematic later on.

## Twins

Twins provide a unique opportunity for a fantastic natural experiment to try to tease apart the effects of upbringing and genetics, and it's no surprise that epigenetic research has used populations of twins to help understand some of what's going on.

Many twin studies use the two different types of twin: identical and non-identical. Identical twins share 100 per cent of their genes: 'They're basically clones,' says Cecil. Genetically they are completely the same. Non-identical twins share about 50 per cent of their genes, the same as any other sibling, but they've grown prenatally in the same *in-uterine* environment. Comparing the occurrence of something in identical twins versus non-identical twins gives us an estimate of how heritable something is: how much any difference is likely to be due to genetics versus environment.

'When you do this with DNA methylation or epigenetic patterns,' says Cecil, 'you see that indeed identical twins not only look more alike, but are also more alike in terms of their epigenetic patterns, compared to non-identical twins.' So identical twins will also have more similar gene expression than non-identical twins.

That's not the whole story, though. Even more interesting, to me, is that even though identical twins start off more alike in their epigenetic patterns, as they grow older these epigenetic patterns grow more and more dissimilar.

'It's really interesting,' agrees Cecil. 'As they grow older, identical twins start to look less alike, and develop their own personalities more – become different. It seems that biologically that's the case as well. It's been proposed that, as they grow older, identical twins get exposed to more and more different things. They might make different friends and go to different schools, maybe one starts smoking and the other one doesn't . . . All of these accumulating new experiences might be helping to shape their biology in a different way, to adapt to their own set of circumstances.'

This was one of the first clues that the environment might be important to epigenetic mechanisms. Since then there's been a huge interest in how the environment affects epigenetics, because it might help us understand, among other things, the origins of health and disease.

'We know, of course, that the environment is really important in affecting the risk of disease,' Cecil explains, 'and epigenetics might be a key mechanism through which this happens. I think what's really interesting about this is that it's not just direct physical exposures that can affect the epigenome. We know that smoking has a very strong effect: whether you look at people who smoke, or babies who have been exposed to prenatal smoking, they show the same sort of pattern. We know certain things are very clearly associated – that diet is very important for methylation, so for example folate, which is incredibly important in pregnancy, is actually intrinsically involved in methylation. Even drug abuse – all of these things can have an effect on the epigenome.'

It isn't only direct physical exposures that can have an influence, but also social exposures. 'That's something I think that's particularly relevant to our understanding of child development and mental health problems as well,' says Cecil. 'We know that factors such as early-life stress, maltreatment, deprivation, are very much predictive of later mental and physical health problems. Epigenetics is something that has revolutionized our understanding of how that might actually work, in showing how what you experience in early-life parenting and social conditions can have a very strong biological impact. It's not just things that you are directly exposed to.'

## Foster-caring rats

A particularly elegant series of experiments with rats illustrated the importance of experiences in shaping genetic expression. In the 1990s Professor Michael Meaney and his colleagues got interested in the effect of early rearing conditions or early parenting on a rat pup's biology and later development, and performed some studies. They're among Charlotte Cecil's favourites, and she talked me through why. 'They noticed that rats, a bit similarly to humans really, vary in the quality of their parenting: they show different parenting behaviours.' Meaney and colleagues measured this by observing licking and grooming behaviours. 'So the idea is that when the pups are born, the more licking and grooming of the pup takes place, the more you have a sort of proxy of caring, warm parenting. Most rats are somewhere in the middle, and then you have these more extreme cases of very, very high licking and nurturing parenting, versus very low levels, which might be considered a proxy of neglectful behaviour.' The researchers focused on these two more extreme ends of high and low licking and grooming. They looked at the pups from these different parents, and compared levels of methylation of a very important gene called the glucocorticoid receptor, or GR for short.

'This is a really important gene,' explained Cecil, 'because it's heavily involved in stress response. It's involved in determining levels of circulating stress hormones – cortisol, for example – and it's very involved in the HPA axis.[2] It also has been shown in humans to be kind of a candidate gene for mental health – stress-related disorders such as depression and PTSD.' The researchers found that rat pups in the high-licking and -grooming group had low levels of methylation of this gene. This meant the gene was easily accessible, not covered by a methyl blanket, and was functioning optimally. This in turn led to more receptors for this neurotransmitter developing in the hippocampus, a brain area important for stress and memory. When the pups were exposed to future stressors, they showed greater resilience, lower levels of circulating stress hormones and fewer stress behaviours.

When the researchers looked at the lower-licking and -grooming group these pups exhibited the opposite pattern: their glucocorticoid receptors showed high levels of methylation. This blocked access to the gene, which meant that the gene was read less and was less active. This resulted in fewer of these important receptors in the brain, and when exposed to future stressors these pups showed higher levels of stress hormones that lingered for longer. Behaviourally, they showed higher anxiety and fearful responses.

The researchers then went one step further. They cross-fostered rat pups from high-licking and -grooming mums to low-licking and -grooming ones, and vice versa. Even for rats who were genetic offspring of the mums who had low licking and grooming, it was found, if they were fostered by a rat that licked and groomed them a lot, then they showed less stress response and higher levels of GR gene activation. They also went on to be high-licking and -grooming mums themselves when they had pups. Foster care helped.

'I think this was really an important study,' says Cecil. 'It was one of the first cases to show that what we might consider as disordered outcomes, such as anxiety, can actually be traced to potential adaptations to the environment via epigenetic regulation of important genes for stress response.

'It was the first study to show that social exposure such as parenting behaviour could have a measurable impact on regulation of a gene important for stress, and, at the same time, that this regulation of the gene could have a measurable response on future behaviour and future response to stressors.'

The idea that epigenetics is how we adapt our DNA expression to our environment is a very helpful one. It might seem like a negative consequence for the rat pups of the low-licking and -grooming mums-to-be showing anxious behaviours, but Cecil disagrees: 'It's always good to think about what kind of adaptive advantages some of the things might actually bring when you think of it in the context of quite a stressful set of situations.' Even consequences which might look unhelpful could be helpful in certain circumstances. In the case of the rat pups, 'You can think of it instead as a very adaptive response of pups who can't rely on their mother for protection and need to be more vigilant – for example, to potential threats in the environment, and maybe wanting to be less exploratory because of potential dangers.'

## Limits and uses

Epigenetics is helping us to colour in some of the gaps in our knowledge about how nature and nurture interact, but there remain several unanswered questions. A vast number of epigenetic mechanisms are not understood, and we are yet to establish the difference in effect between different extents, severities and timings of environmental influence. Epigenetic mechanisms are dynamic, and occurring all the time as part of normal development. The more we understand about them, the more we might also understand how to target adaptations which become problematic later on, and whether we should aim to do this or not. There is some limited evidence suggesting that talking therapies as well as pharmacological interventions can change our epigenetic mechanisms, and how this happens is again of great interest.

Charlotte Cecil sees it in terms of short-, medium- and long-term implications. 'In the short term it really just helps inform existing models of how experiences get under the skin, how nature and nurture come together. In the medium term, as findings start to be replicated more, we might be able to use epigenetics as a way of finding biomarkers: biological indications for exposure, risk, or diagnosis and prognosis. Eventually, longer-term, there might even be potential to intervene at the epigenetic level.'

This longer-term implication in particular needs careful thinking through. Genes don't work in isolation, and targeting one particular gene and changing its methylation levels might have downstream effects that are unhelpful. As Cecil explains, 'If these changes are in place to help an organism to adapt to its environment, how is that going to play out?'

Ultimately, epigenetic mechanisms are only one type of process involved in the hugely complex orchestra of interactions that are going on continually within our bodies, and we don't fully understand them yet. It is tempting to think of epigenetics as a deterministic framework, and because some of the animal studies have suggested that environmental events can alter the epigenome of not only offspring, but also offspring's offspring, there has been speculation that this will be the same for humans. This is still a massively controversial area, and the evidence in humans just isn't there yet. Before we all start worrying about whether our grandparents smoked or drank too much, Cecil urges caution in jumping to conclusions. 'We know that the vast majority of epigenetic marks actually get reset at the time of conception,' she explains. 'And although maybe some might be able to survive this major erasure, we don't know enough at the moment. One thing to say about this, too, is that actually you can potentially affect three generations with the same environmental exposure. Take, for example, a mum who is pregnant, with, say, a baby girl. Say the mum is exposed to a very severe stressor that could influence her epigenome. It could potentially also influence the baby's epigenome because it is obtaining signals from her, and it could affect the baby's germ cells, from which ovaries are formed, and which contain information for the next generation. That's not the same as saying something that my ancestor experienced that I haven't been exposed to is being passed down to me.' To really tease apart this question we need studies to look across four generations in a family, something that is very hard to do.

Cecil is interested as well in trying to understand how pre- and post-natal experiences influence the development of mental-health problems through epigenetics. She is especially interested in co-morbidity (co-occurrence) of physical and mental-health problems, wondering if there are common biological pathways involved. She identifies three main bodily systems as being involved and interconnected: inflammation, stress response and neurodevelopment.

'These things actually seem to be incredibly interlinked, and have a feedback system. If there's something that seems to disrupt these systems, you could see if it might result in an increased risk of mental and physical health problems too. It's still a little bit of an under-researched area – the mind–body connection and all that. But I think it's something that's starting to get a little bit more awareness and attention.'

## Levels of explanation

Epigenetics provides us with a biological explanation of the theory that nature and nurture are both influential. It makes me think of the framework that the neuroscientist and physiologist David Marr came up with: the idea that we can try to understand complex systems on different levels of explanation.[3] The complex system might be the brain, a computer, human behaviour in a group, or in this case the interaction of nature and nurture on human development.

Marr's three levels of analysis are drawn below:

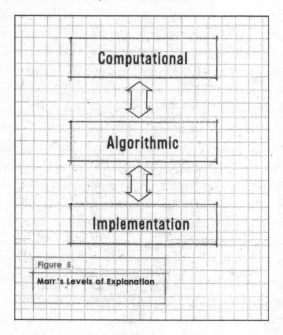

Figure 5. Marr's Levels of Explanation

The computational level of analysis describes and specifies the problems.

The algorithmic describes how the problems can be solved.

The implementational level involves the physical mechanism and its organisation in which the computation happens.

So, for example, the computational level could be seen as the theoretical understanding that there are influences on our development from both nature and nurture. The algorithmic level describes the idea that there is a gene x environment interaction, and the implementational level is the detail of how DNA methylation does this for a particular gene in response to a particular environmental influence. Epigenetics has coloured in a whole new level of explanation for us, filling in the blanks and giving a biological 'how to' for an idea that seemed obvious but had no tangible explanation.

These different levels of explanation can apply to ideas about mind–body links too, and help us to sidestep the classic arguments that we can sometimes get stuck in. Another pair of polarised positions is the idea of a conflict between biological models of mental illness and more social and psychological explanations. In reality, professionals often acknowledge that depression, for example, involves changes in brain chemistry as well as changes in thinking style, not just one or the other. Research into the effects of talking therapies on brain chemistry has found, more and more, that there can be biological consequences of psychological interventions. When we look at this through the framework of Marr's levels of analysis, it isn't surprising that there are both psychological and biological changes. We can see talking therapy as being on the algorithmic level, and medication as being on the implementational level, with the computational level being the subjective sadness that the person feels, and the lack of functioning they have while they feel so bad. The lens through which we look at the problem, influenced by our professional and personal background, colours which level of analysis we might be more interested in.

The nature–nurture debate, and the idea of connections between the mind and the body, both have ancient roots stretching far back through the centuries, and the continual discovery of new biological mechanisms behind these ideas is hugely satisfying. For me, with my teenage years spent in a town where mind—body connections were talked about with certainty, and with memories of reading with total delight, up in my attic bedroom, science books that seemed to validate this,[4] it feels good that some of the influences that got me set onto my trajectory of study are still highly relevant and very exciting.

## 5

# Learning to see

## The mystery of babies

It's really hard to work out what is going on for babies. Sometimes they look like they are the wisest beings on the planet, thinking profound thoughts that they just can't communicate. Sometimes they look entirely gormless, like someone drunk and half-asleep in a brightly lit Tube carriage, unable to focus their eyes on anything. We can't just ask them what they are experiencing, or we can, but they don't reply. We can't do the standard eye test: 'Is this one clearer, or this one? The green circle, or the red?'

It's not even easy to watch babies and draw conclusions from what they are doing, because they can't do that much. Their ability to move about, to lift and move their head, to interact with the world by picking things up or going towards or away from them: all of these take time to develop. So it's hard to draw conclusions from what they are up to, because mostly they are just lying there, sleeping, crying or defecating.

Or watching. Babies do watch, especially the faces of the people holding them. It turns out that as babies we love looking at faces. And the distance that we first begin to focus on is about the same distance from our face to the face of the person holding us.

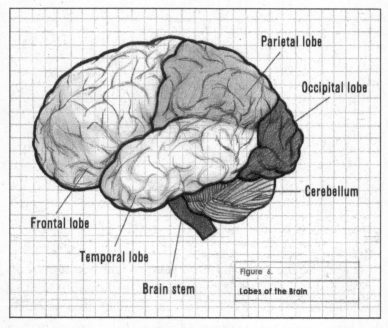

Figure 6. Lobes of the Brain

## *In the beginning*

When we are first born we come into a world full of moving and changing patterns of light, dramatically different from the darkness of the womb. Our eyes and brains need to develop an ability to focus, to distinguish fine detail, and to make sense of the patterns which are suddenly available to us. Imagine spending nine months in the dark and then emerging into the world: no wonder babies sometimes look a bit mind-blown by mobiles dangling from their cots.

The part of the brain most involved in sight is the visual cortex, located at the very back of the brain, in the occipital lobe. Information from each eye is transferred to the occipital lobe via the optic nerve, with information from the left and right visual fields swapping over, so that the left visual field information is sent to the right side of the brain,

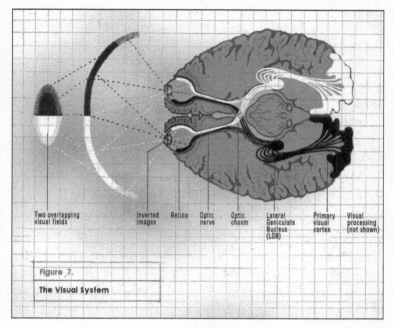

Figure 7. The Visual System

and the right visual field information is sent to the left. What we see in each visual field gets integrated by the brain so that information from both sides is taken into account.

## Cunning experiments

Researchers have spent considerable time trying to work out just what we see as young babies, and when vision improves. This has resulted in some seriously creative experiments, which are only more impressive when you consider that the participants are very young infants, a pretty hard population to recruit and to keep happy. These experiments tend to happen in baby labs, which are places where parents bring their babies to be experimented on. There was one of these at the university where I studied as an undergraduate, and I remember seeing babies and toddlers being carried in, up the stairs or through the tea area among the clusters of students.

Two of the most popular ways of working out what babies are experiencing are called the preference technique and the habituation technique.

In the preference technique, you basically show a baby two different things, and measure whether they are always looking at one of the things for longer than the other. A consistent preference for one of the things suggests they can tell the difference between them. You need to show the babies the objects in a variety of positions, though, to make sure they don't, for example, just have a preference for looking left or right.

Habituation involves showing babies the same thing again and again and again. If you do this, the baby tends to spend less time looking at it the more they have seen it. Researchers interpret this as the baby getting bored of it, because it has remembered more and more of it, the more it sees it. They suggest that babies getting bored like this suggests they have some form of visual memory – some ability to remember some of what they see.

Habituation followed by showing the baby a new thing can be used to work out whether babies can spot a difference between the old thing they are bored of looking at, and a new thing they haven't seen before. If the baby looks for longer at the new thing, it suggests that the baby can tell it is different from the thing they've been habituated to. A neat version of this is to habituate a baby to one thing, then present the baby with two things at the same time to look at: the old thing and something new. Seeing which one the baby looks at for longer should help us work out whether the baby can tell a difference. Again, you have to vary the position of the object, to make sure the baby doesn't just have a preference for looking in one direction.

Another variant of this sort of experiment is to condition a baby to one stimulus and not another: presenting the baby with a particular sound, for example, at the same time as presenting it with something rewarding to look at, like a short animated video clip. This conditions the baby to turn their head towards the sound.

Scientists have even got around the problem of really young babies not being able to move their head much. When we are really little, our heads are just too heavy to be supported by our necks. Really little babies need their heads supported or else they loll about. So instead of tracking where the babies' heads are pointing, scientists track their eye movements to see what they look at.

It doesn't stop there. We can also measure other physical reactions that the baby is having. Monitoring heart rate means we can see if it slows down, stays the same or speeds up. Scientists tend to relate a slower heart rate to concentration, and a faster heart rate to surprise or fear. Whatever the change means, a consistent difference in heart rate when a baby is looking at different things suggests they can see some difference.

There are even techniques to measure what is happening in a baby's brain. EEG (electroencephalogram) is the least costly. It does involve covering the baby's head in monitoring electrodes, but tiny caps have been designed to allow this to be done relatively quickly and without causing the baby too much fuss. It doesn't hurt at all, or have any side-effects, and it means that the electrical impulses inside the brain can be picked up. The main downside of EEG recording is that it is really hard to tell where in the brain the electrical signals are. It tells us very little about which bits of the brain are active, but it can tell us in general terms that activity is increasing or decreasing.

The most popular type of brain scanning, fMRI (functional magnetic resonance imaging), can be used on babies too. fMRI allows brain activity to be measured while someone is in a brain scanner. As long as the person keeps fairly still, they can have their brain scanned while they are doing simple tasks or being asked to think about certain things. The scans measure the amount of oxygenated and deoxygenated blood in the different brain areas, and the working assumption is that levels of blood oxygen change in areas of the brain which are active.

Slightly problematically, the results of fMRI in very young babies seem a little different to the usual patterns of activity in adult brains, meaning the data is even harder to interpret than usual. fMRI produces beautiful images of the brain, which are usually generated in different colours to show the areas where there are higher levels of blood oxygen. These rainbow-coloured images can make it seem as though our brains are lighting up in response to different stimuli. In reality, we still don't know enough about what the blood oxygen signal means, and it is definitely worth exercising caution as we draw conclusions. The excitement of being able to take a glimpse of a brain in action is fairly contagious, though, and advances in scanner technology have meant quieter scanners which can be used on babies without scaring them, and which have good resolution even in short scanning times to allow for babies wriggling about.

## Babies can't see very much

One thing we have learned from all these sophisticated experimental procedures is that newborn babies have very limited vision. It's enough to function as a baby, but not to win any prizes in an eye test. The level of detail we can see when we are newborn is about a thirtieth of what perfect adult eyes can see.[1] This improves pretty quickly, but for the first few months we are seeing the world as a blur. Soft focus, Vaseline on the lens.

As newborns, we also can't control what we focus on very well. We can't see small details but we can see larger objects, as long as they are far enough away from us. A newborn can see about eight to twelve inches away from them, about a good distance to see the face of the person holding them. By six months they develop an ability to see much further, as far as fifteen feet.[2]

Newborns also can't see many colours. If you've ever wondered why a lot of baby books for very small children are in black and white, that's why. As newborns we can tell the difference between black and white, and between red and white, but not between white and other colours.[3] At about two months old[4] we develop the ability to tell other colours apart from white, but until then we have a very limited rainbow.

We also have a limited ability to track moving objects. Our eye muscles don't allow us to follow something smoothly: instead our visual scanning jumps about jerkily. At about one month old we can focus on a small number of features in our surroundings, then at about two months old we can look around more thoroughly. By six months we have more of a smooth scanning ability. Even this poor ability to look around us early on might perversely have some advantage. It might be that having such a narrow range of things to look at and focus on means we really focus on the important things, like the faces of our main caregivers.[5]

We might be rubbish at looking around and focusing but, despite this, at two days old our ability to spot the difference between patterned and unpatterned shapes is surprisingly good. In the 1960s Robert Fantz, a developmental psychologist, showed two-day-old babies pictures of stripes, checks or bull's-eyes, as well as plain discs and squares. The babies seemed to prefer looking at the patterns over the plain pictures.[6]

## Patterns

In more studies relentlessly showing babies things to look at, one- and two-month-olds were given patterns which had identical amounts of light and dark areas in them, but different levels of complexity of pattern. The one-month-olds liked looking at the simpler patterns, and the two-month-olds liked the more complicated ones.[7]

Babies under one week old were shown patterns that either did or didn't have curved edges.[8] They preferred looking at the ones with the curve, adding to the body of evidence that newborns can also discriminate between shapes, line orientations and angles.

A classic study from the 1960s[9] showed infants three different stimuli: one that looked like a face; one that had all the same elements but scrambled up; and one that was blank. All ages preferred looking at the face, then the scrambled face, and lastly the blank stimuli. Why? Perhaps infants just like more complex images. In the 1980s a study tried presenting one- and two-month infants with natural faces, symmetrically scrambled faces and asymmetrically scrambled faces.[10] One-month-olds looked at them all for the same length of time, but two-month-olds looked at the natural faces for longer. Other studies have shown newborn babies faces and non-faces moving across their visual field. Initial studies suggested babies track the faces more than non-faces, and that after three months they prefer looking at static faces rather than non-faces.[11, 12] Not all studies show this clearly, though, and there is still a lack of consensus around a lot of what we know about babies' perception of faces.

## What we don't know

We don't know why there are different results for still and moving faces in very young and older babies. It does seem that tiny babies somehow prefer faces, but we aren't sure what it is they like about them, or how it is that they distinguish them from other shapes. Some researchers have proposed a specific face-processing system in our brain responsible for the preference for faces that tiny babies show.[13] Others propose that there are two different systems in the brain related to vision – one involved with vision for action, and one involved in object recognition – and they relate the different findings for moving and static faces to the employment of these two different systems.[14] We don't conclusively know.

## What we do know

Despite all this facial controversy, we are pretty sure that faces are particularly important to us from a young age. Even very young infants can recognise and learn faces very quickly. Walton and colleagues[15] got one- to four-day-old infants to look at videotapes of their mother and a similar-looking woman. Babies could control who was on the screen by sucking a teat. Babies sucked more for the images of their mother (except for one baby which didn't – that poor mum!).

As for how the infants recognise faces, it seems the edges of the face are important. When the edges of faces are covered with a scarf, babies can't recognise their mums any more[16] from a picture alone.

It makes sense that faces should be so important: since babies are so vulnerable, and so reliant on their primary caregivers, it pays for them to be able to recognise them. Much controversy[17] has surrounded the literature on whether babies not only recognise but also imitate familiar faces. Experiments from the 1970s[18] suggested that even newborns could imitate adult faces, but the consensus now is that those studies were too optimistic, and that in fact what they captured was babies sticking their tongues out in excitement at pretty much most new facial expressions.[19] Imitation of facial expressions does occur, but not reliably until some months later. This isn't to say there isn't any capacity for social reciprocity of gesture, though, as *in utero* studies of twins show some capability for this even before birth.[20]

## 3D or not 3D? That is the question. . .

A classic experiment from the 1960s tried to work out whether babies could perceive depth. It's hard to do this, because parents of babies are typically a bit loath to let them climb up high or be dangled over a precipice to see how they react. Nonetheless, Gibson and Walk[21] came up with an ingenious table-like bit of kit to test whether babies would

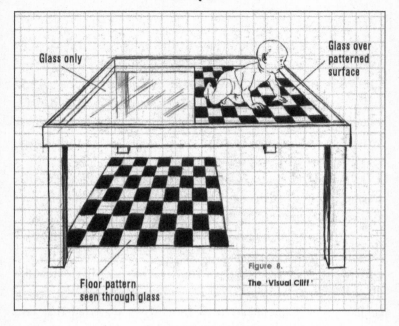

Figure 8. The 'Visual Cliff'

crawl over a glass surface with a drop underneath it, or whether they would only stay on the side without the illusion of a drop. They got a range of babies, aged between six and fourteen months old, popped them down in the middle of the table, and watched which way they crawled. The infants crawled over the 'shallow' but not the 'deep' side. This was taken as evidence that from six months infants can register depth.

Most five-month-old babies can't crawl, but if you place five-month-olds and nine-month-olds directly onto either the shallow or deep sides of the same apparatus, and measure their heart rate, you can see differences.[22] The nine-month-olds' hearts beat faster when they are on the deep side, as you might expect if they could see a difference and felt scared. Weirdly, the five-month-olds' hearts beat slower on the deep side. So they did notice a difference, but they showed no evidence of fear.

This could be to do with experience. Five-month-olds typically can't crawl, so won't have had experience of getting near an edge or falling down a step. To test the idea that experience plays a part in this, some researchers[23] got infants of a similar age, some of whom could crawl and some of whom hadn't learned to crawl yet. Infants who could crawl had increased heart rate when placed on the deep side of the table. Infants of a similar age but who couldn't crawl had a decreased heart rate on the deep side. So far, so similar to the previous experiment. Then, the researchers gave the infants who hadn't yet learned to crawl several hours in a wheeled walker. They let them wheel about to their heart's content and get more experience of moving about in the physical world. Then they tested them again on the deep side. Their heart rate increased. So it suggests that getting experience of moving about the world prompts us to learn that the deep side might be dangerous.

### The effect of experience

It isn't only humans who learn from experience. A lot of animal studies have investigated the effect of experience on the development of our visual systems. In the 1950s one experiment[24] tried raising rats in the dark, and then assessing their reactions on a visual cliff like the one the babies above were presented with. Even those rats raised without experience of seeing depth favoured the shallow end over the deep end, which goes against the infant findings above and several other animal studies. The grisly kitten experiment described in Chapter 3 shows how, in contrast, some visual experience is really important to visual development. Similarly, kittens raised with only an environment of either vertical stripes or horizontal stripes become visually impaired.[25] They never develop the brain cells they need to be able to see stripes which are in the other orientation, and their ability to see the world is compromised.

## Seeing things

We don't all have the same experiences of seeing things. We might be short-sighted or long-sighted; we might be colour-blind, or have specific difficulties recognising faces or objects; or we might see things that other people don't see. Visual hallucinations are a symptom of psychosis, or loss of contact from reality, but in fact unusual experiences like seeing things or hearing voices are way more common than most people think. Children and adults might experience sensations that others don't, and it doesn't signify madness or dysfunction. It only becomes a problem if it's upsetting or it gets in the way of day-to-day life.

Even without experiencing anything unusual, we still perceive the same situation in different ways.[26] The way we view the world used to be thought of as our basic senses feeding into our brain to be processed at a higher level. Current thinking is that our higher-level brain processes are involved much earlier on, and that what we expect modifies what we see. People who have been put in a bad mood who are shown different expressions of faces on either side of their visual field are more likely to notice the sad or grumpy-looking face than the happy or neutral one.[27] In a similar type of experiment, two images are presented on either side of the person's vision, but this time one image of a person and one image of a house: people who had already heard some negative gossip about a person were more likely to notice their face in their visual field than those who had either heard nothing or heard positive or neutral gossip.[28]

These are neat experiments and they also reveal just how powerful the human brain is. Part of learning to see is the honing of the machinery we have available to us to do this, and the development of our brains to enable interpretation of the information. Another part is related to our wider brain development and knowledge of our social world. When we expect something to happen, or when we are primed to see something, we are more likely to notice it. This might be protective, when we think about our evolutionary roots as potential lunch for a sabre-toothed tiger and therefore the need to be wary of our environment and possible threats, but it is much less useful in today's relatively safe environment. It might instead lead us to become trapped in a vicious cycle of our mind's own creation – expecting threat and therefore perceiving threat; feeling blue, and therefore being extra-sensitive to noticing sad things. In the words of Mark Twain, 'You cannot depend on your eyes when your imagination is out of focus.' On the flip side, if we are aware of this, we might also be able to make a conscious effort to look out for the opposite: to seek out evidence to the contrary of our expectations, or to train ourselves to have more optimistic ones.

# 6

# All about the chat

Most of us knew the grammatical rules of our language by the age of five, enough to generate unlimited sentences. Whether we could have explained the grammar is another matter, but it's likely we knew enough to speak it well, whatever the language, whatever our cultural background and whatever part of the world we were from.

Children meet their milestones for language acquisition at a very similar rate across the world, suggesting a common way of learning language and providing a helpful norm for thinking about the times when things go awry. Although there is a great deal of knowledge about how we acquire language, there is still a raging debate between two schools of thought on how best to explain it.

The stages that we go through as we develop an ability to use language are widely agreed on. Despite coming from different backgrounds, locations, upbringings, ethnicities and sexes, most children follow the same milestones in acquiring language. Aside from crying, which can certainly be classed as a communication, by six weeks old babies are already cooing in vowel-like sounds. By two to four months they are gurgling too, playing with sounds that they can make. By four months they begin to be able to generate consonants like 'g' and 'k'. By six to eight months they are babbling, which is the technical term for putting vowels and consonants together. This experiment with articulation continues and becomes more and more repetitive, and then increasingly complex, until between twelve and eighteen months children begin to say single words, but in a way which often seems to indicate a more complex meaning. Between eighteen and twenty-four months infants are able to put two words together, and from about thirty months old they can put multiple words together.

We are born with a predisposition to perceive the sounds of language. Children as young as four days old can tell the difference between things said in their mother's language and things said in another language.[1] One clever experiment studied infants who were only aged nought to three days old. Researchers visited the babies in the maternity hospital, using special soft cloth caps to hold sensors in place which used near infrared spectroscopy (NIRS) to measure brain activation responses to hearing recordings of different languages. Researchers played the babies, who were born to monolingual English parents, recordings of English, of Tagalog (a Filipino language) and of backwards English and Tagalog (to provide a nonsensical control condition). Babies' brain activation responded differently when played their native language as compared with being played their non-native language and the nonsense language. The language they heard *in utero* seems to have affected their sensitivity to their mother tongue before they were even born.

## How do we learn language?

The steps we all go through to get from silent observer to linguistically competent participant in the world are known, but how does it work? Dr Froso Agyri from Birkbeck University took the time to explain to me some of the mechanisms involved. 'Language acquisition process is relatively fast,' she says, 'considering the complexity of the grammatical rules children acquire. Children pass through different linguistic stages. They begin by babbling; then they produce their first words; after a few months they begin to combine one or two words together into sentences, and then they start using more complex sentences.'

It isn't just a case of memorising. Babies and young children are never explicitly sat down and instructed in the grammar of their language: instead, they manage to extract grammatical rules from the world around them. Their capacity to hear different sounds is huge, in fact greater than that of grown-ups. Until about six months infants can respond to contrasts in phonemes[2] in a huge range of human languages, not only that of their mother tongue. This ability disappears, though, making it harder to learn languages, to the extent that we can speak them like a native, if we haven't had exposure to the phonemes from a very early age. A good example of this is how hard it is for Japanese children to hear the contrast between 'r' and 'l' past about six months old, whereas English-speaking infants can.[3]

All infants babble, and as they get older their babbling takes on the characteristics of their language more and more. Deaf children exposed to sign languages from birth produce manual babbling, using signs and gestures they have been exposed to. Babbling has been thought of as a universal first stage in language acquisition shaped by whatever input babies are receiving from the world around them.[4]

From babbling, babies of about nine to twelve months old start to produce words. By fifteen months old most infants have about fifteen words in their vocabulary. The appetite for language is voracious, and even more conservative estimates of language acquisition suggest that the average eighteen-month-old learns one to two words a day, an average four-year-old about four words a day, and a seven-year-old as many as twenty words a day.[5] Of course, all children's learning abilities are individual, so different children will have different vocabulary.

Around the age of one year old children seem to realise that sounds are linked to meanings. Children vary hugely in when they produce their first word, but it seems to be linked to this understanding that the words are communicative. Many children use single words to indicate more complex meanings. Agyri gives an example: 'When the child says "down" he may be asking to be put down, or he may be talking about a toy that has fallen down.' Studies suggest children of this age understand more than they are able to say – for example, they are able to tell the difference between 'What is the apple hitting?' and 'What hit the apple?'[6]

The next stage children go through, usually at about eighteen to twenty-four months, is the two-word stage, when they start combining words in two-word groupings, like 'dirty cat', 'Daddy play', 'choco mine'.

Agyri describes the next stage as producing sentences that are a bit like telegrams, 'including only the words needed for basic comprehension'. This stage is known as the telegraphic stage, and children skip out the linking words that we need for a proper sentence: articles, auxilliary verbs, endings, they are all out of the window, yet the utterances still make some sense. 'Tilly want hat', 'That a big one', 'Daddy make dinner', 'Mummy go sleep'. There is some basic meaning in all of these, and also some knowledge of correct word order.

Between two years and six months and three years old, language development is really fast. 'By the age of three, the basic grammatical principles of sentence formation are acquired,' explains Agyri. 'Children have begun to use and understand complex structures and sentences.' The relative complexity of the sentences children can understand and generate increases as children get older, and by four most of them have got the grasp of most adult grammar.

Considering this is all done without much explicit teaching, the speed at which children become masters of their own language is phenomenal. The process seems to involve some over-learning of rules and generalisations.

'Children seem to make specific types of errors,' says Agyri. 'English-speaking children, while acquiring morphology, may incorrectly overgeneralize the regular past tense ending -ed or the plural ending -s.' They would use -ed for everything, without making exceptions for irregular examples like 'went' – instead saying 'goed' – or for plurals such as 'fish', instead saying 'fishes'.

Children can also over-extend and under-extend word meanings. 'In the first case,' explains Agyri, 'they would use "cat" to refer to any four-legged creature, whereas, in the second case, they would use cat to mean her pet and no other cat.' Similarly, all round things might be called ball, and all furry things might be called bunny. Agyri thinks there is a reason for this: 'Children can learn as many words as they do because they are guided by such general principles. Children learn about fourteen words a day until the age of six, about 5,000 words a year!'

## What happens when we learn a second language as an adult?

The process of learning our own language as a child happens without us consciously trying. By contrast, trying to pick up a new language is usually a conscious choice. 'Individuals who learn a second language as adults often find it a challenge,' said Agyri, 'which does not resemble the experience of learning their first language. Memorization, conscious attention, and possibly intense study is required to become native-like in a second language.' Frustratingly, when we learn a second language as an adult we usually can't reach a stage where we sound just like a native speaker and don't make at least some grammatical errors, but the younger we are exposed to a second language, the more likely we are to be able to learn it to a native-like level.[7] Agyri is encouraging, though: 'Although age is an important factor in achieving native-like competence as a second language, adults can still acquire the grammar of the second language.' It does seem that language-learning abilities get less as we get older, and some of this might be because we have missed so-called 'sensitive periods' in development where our brains were particularly good at learning aspects of language, for example the greater sensitivity to different sounds that young infants have, compared to adults.

Even though we might not do as well learning a language later in life, there is some evidence that we still learn in the same way. 'Second-language learners pass through the same developmental stages while constructing their second-language grammar as children do during first-language acquisition,' says Agyri. 'In other words, the acquisition process of a second language does not seem to be fundamentally different from first-language acquisition.'

## Is there an innate capacity for language?

Given how children the world over seem to learn language similarly, it does seem that we must have an innate capacity for learning language. This remains a controversial point, though, and linguists and philosophers alike argue over it. There are two main theoretical poles to the debate: behaviourism and nativism.

A behaviourist point of view would argue that children learn language by copying, being rewarded when they get it right, and corrected when they get it wrong. This point of view was really popular in the 1940s and 1950s in particular, and is still held by some language theorists today, although the evidence around how children learn language doesn't support the idea that they copy or are explicitly rewarded or corrected by their caregivers. Even if children make grammatical errors, parents and carers tend to correct these less if the sentence makes sense in terms of content. Children also can't be learning everything by copying, because they make mistakes that they wouldn't have heard around them, like the over-generalisation of past tense or plural rules.

The nativist position is represented by Noam Chomsky, who proposed instead that all children are born with some knowledge about general language rules. This would be knowledge about the form of a language, rather than the details, but would aid language acquisition.

Agyri sees sense in this: 'The ease, rapidity, uniformity and universality of the language-acquisition process suggest the innateness of the language faculty, i.e. that humans are born with an innate template or blueprint for language: Universal Grammar. Universal Grammar is not like the grammar of English or Greek,' she goes on: 'rather it represents the general rules that govern human languages. Children actively construct their grammars based on the linguistic input, and are supported and guided in this task by Universal Grammar.' She quotes Chomsky, who said, 'We are designed to walk ... that we are taught to walk is impossible. And pretty much the same is true of language. Nobody is taught language. In fact, you can't prevent the child from learning it.'[8]

## Is there a 'language learning bit' in the brain?

If there is such a thing as a Universal Grammar that gives us a predisposition for language learning, then where in our brains is this knowledge? There is again a controversy here, over whether there is a general learning mechanism in our brains that is involved in this, or whether there is a specialised part of the brain which acts as an 'acquisition device'. Part of this controversy can be seen as stemming from a more philosophical disagreement about the fundamental nature of language. Some linguists see the primary feature of language as being its characteristics as a complex formal system of rules, whereas others think of it primarily in terms of its communicative function. Those who see the grammar structure as primary tend to think there must be a formal acquisition device, and those who prioritise the communicative function tend to think there could be a general learning mechanism. These debates have been going on for centuries, and it's easy to see why: they get right to the heart of the philosophical debate about the nature of language, and philosophical debates are not necessarily designed to be resolved.

## Implications for teaching and learning language

The research and thinking about language development suggests that children really do learn language as long as they are placed in an environment that promotes this. The more access they get to speech and to reading, the more they have the building blocks at their disposal to make sense of these. Children from homes where they are not talked to very much have less opportunity to learn language, and might present with speech delays. Children from homes where more than one language is spoken tend to learn both languages, but might take longer to be able to speak in either language, probably because they are trying to code for two sets of languages in their brains.

## The language of computers

Human-to-human communication is no longer the only language available to us. As the human landscape becomes increasingly digital, communication with computers grows more important. Tom Crick, Professor of Computer Science and Public Policy at Cardiff Metropolitan University, thinks everyone should be taught about the language of computer programming. 'I would say teaching everyone how to program is increasingly a life skill, because it's a digital competency,' he explained to me. 'You don't need to be a programmer to benefit from having learned how to program at some level. Part of it is about equipping you with a skill that is really valuable in the world in which we live.' For Crick, understanding even a small amount of programming gives people more insight into how computers work – 'so they're not just magic black boxes,' he says. 'It gives you a bit of insight into how things like algorithms and computational processes actually affect your everyday life, from the information you read online to job applications, mortgage applications, credit cards ... Actually, you need to understand these things are happening, and are making decisions about us, and they have very little if any human intervention. And, even worse, they capture human biases and prejudices in the same way that humans have them.' The shock that many people feel after election or referendum results come in is one example of how digital media silos can shield us from opposing views, by using algorithms that show us only things they think we will want to see.

For Crick, knowledge is empowerment. 'I think it is a digital competency to understand how computers work, and that to solve a task they have to be programmed by humans,' he says. 'It empowers you to be creative. It gives you the ability to make new things in a digital world.' Crick wants all young people to know that they can program computers and don't need to be passive recipients of technology.

'One of the aims of the new computing curriculum in England is to say, "A high-quality computing education should equip young people to be able to use creativity and computational thinking to understand and change the world,"' he explains. This is ambitious, Crick acknowledges, but he thinks young people should have skills to manipulate the digital world they live in. 'The world is intrinsically and inherently computational,' he goes on. 'In the same way that you teach people about forces and gravity, or you teach people about calculus, you should give them the skills with which to model and understand the world.'

Crick sees learning computer languages as having some overlaps with learning spoken languages, but also key differences. Learning by doing is important, as is 'failing early and failing often', in order to learn from mistakes and understand the structure and key components of the language. Where computer languages differ is in their purpose, and the additional possibilities for vagueness and emotional tone. 'It feels like there is a fundamental difference, in that you are trying to instruct a machine to do something,' explains Crick. 'You have to be syntactically correct much more when you're writing computer code to express the semantics you want to express, as opposed to how we know we can speak grammatically incorrect human languages and still convey the information at some level.' Computer programming leaves no room for sarcasm.

## The human element

Digital languages are a helpful counterpoint to human communication. Human communication lacks the precision of computer programming, but it has additional layers of emotional meaning and interpretation, and a beauty which is hard to capture. Poetry, alliteration, lyrical prose, the way we play with language as a species – all this goes far beyond simple instruction. The way that children go from knowing no language at all to being masters of the spoken word[9] is one of the most mind-blowing of all the developmental trajectories. How some then go on to use those words to affect the emotional states of people all over the world, in plays, novels, poems and song lyrics, is even harder to comprehend. Even in the everyday, the ability to read into nuances of tone, choice of words, even the lengths of gaps in a conversation, how someone else may or may not be feeling about us, is a hugely human skill. It's one that can get us in a pickle when we overdo it, too, and read too much into what people's tone means, or equally when we ignore social cues that something is going awry in our communication. Despite its huge sophistication, in some ways learning our mother tongue itself is just the beginning of developing a massively complex and sensitive capacity for social communication that we use and hone our entire lives.

# 7

# Mind-reading

## The Sally–Anne test

The story goes like this: Sally has a basket and Anne has a box. Sally also has a marble. She hides the marble in her basket while she goes for a walk. While she is gone, Anne takes Sally's marble out of the basket and puts it into her own box. When Sally comes back from her walk, where will she look for her marble?

Hopefully you said that Sally will look in her basket, because that's where she left her marble. If you had a more complex theory, that she'd look in Anne's box because she knows that Anne has kleptomaniac tendencies, then that's something different, but usually adults and children from about the age of four years old understand that Sally looks for the marble where she left it.

Sally and Anne are dolls, and this is an example of a false-belief task, designed by psychologists in the 1980s to see how good children were at understanding what was in someone else's mind. Researchers over the years repeated this experiment again and again, and concluded that because the children under four know that Sally's marble is in the box, they think that everyone does. After four, psychologists thought that children develop an understanding that what's in their mind might not

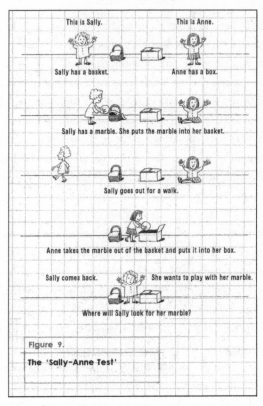

Figure 9. The 'Sally–Anne Test'

be the same as what's in everyone else's mind: that we might not all have the same information at our disposal.

Another variation on the Sally–Anne test uses a well-known brand of sweets that come in a tube. Children are shown the packaging, and then shown what's inside – actually a pencil, and not sweets. The pencil is then put back inside, and the children are asked what someone else would think was inside the tube. Again, under four years old children tend to say 'pencil', seeming to think that other people will know there is a pencil in there because the child knows that.[1] Around the age of four years old there appears to be a conceptual change in children's understanding of what other people can believe.

Psychologists got really excited about this sudden shift, and thought they had pinpointed the age of development of theory of mind – the age where children understand that what's in someone else's mind isn't necessarily the same as what's in their own mind. Which is not to say that all four-year-olds suddenly sit about theorising about the beliefs of others, but more that at about the age of four, children develop an ability to think about their own and other people's mental states: thoughts, intentions, knowledge, beliefs, doubts, pretences . . .

These mental states aren't immediately obvious: we can't see into someone's mind to know what is going on in there, and it takes us some time to appreciate that what's in our heads is possibly different from what's in someone else's head. Just because I am thirsty, it doesn't mean that you are. Having this theory-of-mind capability is incredibly useful to us in the social communities that we live in. We use it to understand and predict the behaviour of people around us. If we didn't have theory of mind, the world would be a much scarier place, and we would be much more likely to get things wrong and offend people or do things they don't want us to do.

For many years everyone was very pleased with this robust and reproducible finding, but it turns out the story is more complex than it first appeared.

## Babies with theory of mind

In the late 1990s an experiment[2] was carried out with fourteen-to-eighteen-month-old babies. They were presented with an adult and two different foodstuffs: cheese crackers and broccoli. The adult acted as if they were revolted by the cheese crackers and thought that the broccoli was delicious. When the toddlers were asked to give the adult one of the foodstuffs, fourteen-month-old infants gave the adults the cheese crackers, and eighteen-month-old infants gave them the broccoli. So the fourteen-month-olds gave the adults what they themselves preferred (none of the children liked the broccoli), whereas the eighteen-month-old children gave the adults what the adults had shown that they liked before.

This was surprising! Eighteen-month-old babies were not supposed to have enough theory-of-mind capability to be able to know to offer broccoli to the adult even though the infant thought it was gross.

So what's going on? Do we actually have more ability as early infants than we think? A whole range of experiments using babies and toddlers suggests that some elements of theory of mind are in place earlier than age four. When infants in their second year of life are shown false-belief tasks that require some kind of spontaneous behaviour as a response, as opposed to requiring them to answer a question about someone else's beliefs, then they seem able to grasp the idea of someone else having a different point of view to them.

These two types of experiments are known as spontaneous-response and elicited-response, with spontaneous responses just being observed by the experimenter, and elicited responses being deliberately asked for. It's hard to have a big conversation with an eighteen-month-old, but you can ask them to point to something. You can also watch where infants look, and measure how long they look at something. Researchers have capitalised on this and measured anticipatory looking, which is where infants look just after a false-belief problem has been shown to them, to see what they expect the agent in the problem to do. So, for example, in the Sally–Anne test, where would the child look just before Sally opens the box or looks in the basket?

The inference is that where the children look is where they expect Sally to look. Researchers also use measurement of gaze to see whether infants reliably look longer at one response to a false-belief test. The inference here is that if they look longer they are surprised, and that this is because of a 'violation of expectation'. So with Sally–Anne, if Sally went to get the marble from the box and the children expected her to get it from the basket, they would look reliably longer than in situations where she went to the basket for it.

A whole range of different false-belief tests have been tried,[3] using toys and real people, and consistently infants in their second year of life seem to be able to understand that someone else can have a different point of view, even if it contradicts reality. Some critics suggest this is more to do with infants learning associations or being surprised by something strange, but the tasks aren't that strange, and the infants don't look longer at unfamiliar pairings of responses and objects if they don't relate to a violation of an expected false belief. For example, even if they watch someone always reaching for a red car and then the person reaches for a yellow car, they will only look longer at this situation if the yellow car was previously visible to the person, suggesting that the person had a preference for red which was now being violated.

So the old idea that four years old is a simple cut-off for theory of mind has been challenged. And it turns out that even the idea that there is a cut-off in understanding at any age is too simplistic.

## More flexible understanding of theory of mind

If theory-of-mind capability was a straightforward capacity that we have or don't have, then we'd expect that the same child would consistently fail or pass false-belief tests. But they don't. The same child might pass some tests and not others.[4] Scores on how many theory-of-mind tests they pass have also been shown to correlate with how high teachers score children on a measure of social ability, leading some researchers to think more about how theory of mind relates with other social skills, rather than seeing it as a purely cognitive problem.

Studies with children over the age of four reveal that there are ways of making the false-belief tests harder, and pushing the age limit where children succeed upwards. Although older children (ages seven to nine) are better than five-year-olds at allowing for differences in opinion between different people, this gets harder if science or morality enters the picture. It seems to become more difficult for children to appreciate that other people can have a different point of view if it's about a scientific 'truth' or about a moral dilemma. Hardly surprising, really, given that as adults we also often find it hard to appreciate that people can have different, but equally valid, views on science or ethics. Adding an emotional component to a false-belief task also makes it harder for children to perform well, and even adults find this tricky, with some studies showing only a 70 per cent success rate in adults. In false-belief tasks which are about objects being in unexpected places, adding more possible locations for the object to be hidden makes them more difficult too.

All of these manipulations are adding in other dimensions – adding more information or other types of processing that are needed. In reality, theory-of-mind capabilities are used alongside other abilities: reading of social cues, understanding of context, processing of other information, registering of emotions – all sorts of extra factors. An ability to process all this information effectively and to inhibit action on some of the information is also important.

This overlaps with the role of executive functioning. Executive functioning describes a range of 'higher-level' cognitive (thinking) abilities that control and regulate other abilities and behaviours. Executive-functioning abilities include the ability to shift and control our attention; to hold information in our working memory and manipulate it; to reason, problem-solve, plan and be flexible in our thinking. These abilities are crucial to our lives as humans, and much of what we do each day involves our executive functions. Theory-of-mind capabilities are significantly linked with executive-functioning abilities. If executive functioning is compromised, then people do less well on theory-of-mind tasks, so there seems to be an overlap.

## Teenagers and theory of mind

Iroise Dumontheil, a researcher at University College London, got interested in how teenagers do on different theory-of-mind tasks. Recent research into teen brains (see Chapter 16) has shown that social areas of the brain are still developing much later than previously thought. Dumontheil and her colleagues designed a series of experiments putting false-belief concepts into a social, communicative framework. In these experiments teens were asked to help a 'director' arrange objects on a modular shelving unit, but certain segments of the unit were covered up from the director's perspective, so they couldn't see everything. Teenagers, the experiment revealed, were worse than adults at telling the director where to put things. They tend to have a more egocentric bias, seeing things more from their own point of view. Theory of mind in this context is still developing.

Theory-of-mind abilities are much more than just performance on the false-belief task, and research seems to be seeing theory-of-mind capability increasingly as a spectrum of performance, rather than an all-or-nothing attribute.

## And still more of a spectrum

The idea of having a spectrum of theory-of-mind abilities fits with a commonly discussed diagnosis, that of Autism Spectrum Disorder (ASD). ASD doesn't only involve theory of mind: it has quite complex criteria, many of which involve early developmental milestones, such as whether we shared toys as children, whether we talked at a usual age . . . One feature of the diagnosis is finding it harder to take other people's point of view emotionally. This can make it much easier to make social gaffes, and it can also leave individuals very vulnerable to being manipulated. ASD is diagnosed more in boys than girls, a phenomenon long suggested to be down to genetic vulnerability to autism, but this is yet to be understood.[5] Whilst autistic-spectrum conditions do seem to be more common in boys than girls, it is worth noting that some part of the low diagnosis rate might be to do with boys' symptoms being more likely to be seen as ASD, and girls' symptoms being seen as something else, like emerging personality disorder, anxiety or depression. The reasons behind this could be linked to wider notions in society of how we describe and understand male and female behaviour.

## Adults and mentalisation

As Renee Baillergeon, a specialist in infant cognition, and her colleagues put it, 'As adults, we routinely interpret others' behaviour in terms of underlying mental states. Thus, we readily understand that Cinderella wants to go to the ball, does not know her fairy godmother will soon arrive to make her dreams come true, and falsely believes she will spend yet another evening mending clothes by the hearth.'

But are we always that good about taking into account differences between what we are thinking and feeling and what other people might be thinking and feeling? We know that we can't read one another's minds, but sometimes we behave as though we think we can. How often do you think you can read what someone is thinking from their facial expression, or from the way they behave with you? How often do you get it wrong? Probably many times a day.

Mentalisation is a relatively new term for an ancient human capacity: our ability to interpret our own and others' actions on the basis of mental activities. 'Seeing ourselves from the outside and others from the inside' is another way of putting it, or 'holding mind in mind'. However we describe it, it's the idea that our internal worlds influence how we act in the external world. It's closely linked to theory of mind, but I think of it as stretching a bit beyond. Where theory of mind is a discrete cognitive ability linked to lots of other cognitive and social abilities, mentalisation is more of a framework for understanding how we inter-relate.

Imagine a typical day. You walk into a coffee shop to get a drink, and the woman behind the counter glares at you. What is your reaction? Do you assume she is having a bad day? Think that you must have offended her yesterday? Or jump to the conclusion that she hates you? Whatever you go for, you are having some kind of thought about her expression: you are making sense of it.

When other people try to make sense of us, it can go both hideously wrong and blissfully right, and of course all the shades in between. If I'm feeling sad and I'm a bit quieter than usual, how nice it could be to have a friend notice and ask if I'm OK. If I'm not sad, just feeling a bit less chatty, and the same person asks the same questions, it doesn't feel the same – it can even be annoying. Like being told to calm down when we're not even that cross, or being told we look really tired when actually we feel fine. It's not so good to be got a bit wrong.

It can lead to really vicious cycles of people jumping to conclusions about one another's beliefs and behaviours. Take Sophie and her mother. Sophie tells her mother about a promotion she's got at work. Sophie's mother feels proud, and also anxious on her daughter's behalf – she wants her to be happy. 'Do you think you'll be all right?' she says. Sophie thinks to herself that her mother doesn't think she can do the job. She gets upset, starts to cry, and her mother interprets this as Sophie feeling unsure. 'You don't have to do it', her mother says, and now Sophie blows a fuse, shouting at her mum for never being supportive. Sophie's mum has no understanding of why, and Sophie has no idea that her mum actually thinks she'd be great in the role.

Or take an example from the clinic room. It's not uncommon to be sitting in a psychology session with a teenager who is saying nothing, tapping his foot like crazy, sitting with his arms crossed and eyes closed. The trap, in this situation, is to presume that I know what they are thinking or feeling. I don't. I don't have a clue. The young person could be cross with me about something, he could be bored, he could be annoyed about an argument he just had on the phone, he could be really sad and worried that he might get upset. His ability to communicate how he is feeling in words might be compromised by his ability to regulate his emotions enough to feel able to speak, or might have been influenced by previous times when he has tried to tell someone how he feels and has come up against misunderstanding. He might have already decided that it's just not worth bothering telling me how he feels, because adults typically don't understand and just bang on about something else irrelevant.

## When mentalisation goes awry

Professor Peter Fonagy, a psychoanalyst and clinical psychologist, and Professor Antony Bateman, a consultant psychiatrist and psychotherapist, have spent a lot of time exploring and defining mentalisation. Professor Fonagy spoke to me about the inevitable fallibility of our mentalising capacities.

'One does not have privileged access to other people's thoughts and feelings,' he explained – 'nor in fact does one have very good access to one's own. You do things and then come back to reflect on it, rather than think through it as you're acting much of the time.' Fonagy thinks it is inevitable that we read ourselves and others wrongly a lot of the time: 'It can go wrong in terms of just being inaccurate – one gets it wrong. Or it can go wrong in more subtle ways, where the ideas are slightly biased, distorted, usually in the direction of being somewhat flattering or more self-serving than they actually should be.'

In relationships with others we find being misunderstood in this way incredibly hard. 'There is an expectation on others that we should be understood, and it's quite painful if people misunderstand or misconceive one's intentions or the reasons why one has done things,' says Fonagy. 'Perhaps we are a little bit more willing to give ourselves the benefit of the doubt in relation to understanding others' actions than we are willing to give others to understand us.'

That sense of someone not 'getting' us is incredibly frustrating, and all the more so if it's someone we really care about, whose opinion we set a lot of store by. We all misunderstand people and are misunderstood by people every day of our lives that we interact with others, but for some people difficulties with mentalising themselves or other people can cause much bigger problems and much deeper distress.

## Mentalisation-based treatment

Mentalisation-based treatment was developed by Fonagy and Bateman in response to one of the most historically difficult-to-treat diagnoses: Borderline Personality Disorder. This diagnosis is characterised by huge and rapid changes in mood, often accompanied by a long-standing sense of emotional emptiness; great difficulty regulating these changing and powerful emotions; difficulties in relating with other people without having huge rows or misunderstandings; and a tendency to act impulsively and in a self-harming way. People experiencing this constellation of symptoms often present to services in an emergency, having taken an overdose or hurt themselves significantly in some other way.

Fonagy thinks the diagnostic label is a poor one (as do many others). He thinks mentalisation-based treatment is 'most appropriate for people who have difficulty in organizing their ideas about how other people think and feel, and about their own experiences organizing their own thoughts and feelings in relation to their own actions.' He acknowledges that 'these individuals sometimes come with a rather inappropriate diagnosis of Borderline Personality Disorder, which is about as inappropriate a term as I can come up with, because I don't think they're on the border of one thing or another. I don't actually think it's probably very much to do with personality: it's probably much more a communication problem. And third, it's probably not really a disorder either. Otherwise,' he adds, tongue in cheek, 'the term is perfect.'

For Fonagy such individuals experience persistent distress and intense internal pain because they are not able to organise their thoughts and feelings about others or themselves in a helpful way. 'They need help from another human being,' he explains, 'who works with them, validates them, clarifies their thoughts and feelings, presents alternative ways perhaps of looking at the world, rather than the rather rigid way they conceive of others, and the presuppositions they have about others' motives.' Fonagy sees this as the way to free up a capacity, 'first of all in the context of that relationship, and then other social relationships, for them to be able to understand others better.'

Mentalisation-based treatment specifically tries to help individuals make sense of how they and others are feeling, by both the patient and the therapist naming their own emotions in relation to interactions. It is crucial that the therapist is curious about what the patient is feeling, instead of presuming that they can tell them. By having experiences of a relationship where feelings are named and understood, and the emotional effects of actions are also named and understood, the therapy aims to help the patient to 'reinstate mentalising' in more situations in their lives. We all might lose the ability to mentalise – to think through the meaning of our and others' actions – when we are in the throes of a large emotional response to something, but hopefully we get it back fairly quickly. For individuals who have difficulty with this, it might be nearly impossible to click back into a state of mind where they are able to think about their own reactions and another person's. Instead, they are left lurching from one emotional response to another, with little sense of solid ground.

Fonagy is clear that he does not see mentalisation-based treatment as a new therapy school, and in fact would be upset if it became this. 'MBT is one of 1,246 different types of psychotherapy, and any psychotherapy is just a box of tricks, a box of tools. I don't believe that there are 1,246 different ways of changing a person. I do not believe that mentalisation-based therapy is the only therapy that works, or even the therapy that works best.' He does believe that it can be helpful as one approach, and in particular for people who tend to be diagnosed with BPD, and the studies suggest that it tends to be effective in reducing self-harm.

'To me therapy is a science,' explains Fonagy. 'It has to be evidence-based; it has to be something that is willing to expose itself to public scrutiny, and face the possibility that some of the people that are being treated in this way are harmed by it rather than helped. There's no effect without side-effect.' The potential harm from mentalisation-based approaches for Borderline-Personality-diagnosed individuals seems much less than from traditional in-patient admissions.

## Attachment

Fonagy and Bateman related the difficulties commonly grouped together in a BPD diagnosis to early childhood experiences. They thought that the underlying common factor was a difficulty with mentalising: trouble for an infant in understanding their own minds and the minds of others. Researchers thought this arose from difficulties early on in the children relating to their parents. Fonagy describes how, if an infant is crying, what it needs is for its caregiver to recognise the emotion it is feeling, reflect this back and name it, but not to get overwhelmed by that emotion themselves. If the caregiver finds it hard to understand and label the child's feelings, the child struggles in turn to learn what these are and to be able to name, understand and control them. If the caregiver gets overwhelmed themselves when the infant shows emotion, then the infant never learns its emotions are not overwhelming. If the infant doesn't learn these things in the first few years, thinks Fonagy, then it will struggle later on to be able to regulate its own emotions and relate to other people.

Fonagy compares learning emotional language to learning any other kind of language: 'Obviously we have to learn this capacity.

Understanding the language of the mind – thoughts and feelings – is just the same as understanding the spoken language, and it's acquired in a social context, usually the context of the family, that is [in the context of] attachment relationships,' he says. The role of our parents and caregivers, our siblings and others around us, is key in shaping this ability. Fonagy sees an overlap with attachment relationships, and goes even further: 'There may be even a deeper thing, which is that attachment relationships might have been hijacked, in a sense, by evolution, to help children and infants learn about minds – that we have mums or dads who are really close to us, who reflect on our thinking and feeling. And maybe we learn about our own thoughts and feelings not by a realization that somehow emerges – you know, "I think, therefore I am" – not something that comes from within, but maybe slightly more accurately, "I'm being thought about by my mum as thinking, so perhaps my thoughts are valid and real."' Fonagy sees our sense of our internal world as developing from an interaction in which we are being constantly recognised and validated by our attachment figure.

While having a good enough attachment relationship is important, Fonagy doesn't think a relationship with one caregiver is the only thing. He sees whole webs of interactions between people around the child as having the potential to be mentalising or not. 'What I'm increasingly convinced about is that it's the system that's mentalising or not mentalising,' he says. 'That's OK: it's access to the system that's ensured by having one or more close mentalising relationships. But ultimately if the system is hostile, no matter how much mentalising you have from one other person, you will have a miserable life. If the system is benign, I think it's a very different story. And some people do not benefit from a benign system until they have validating relationships with one or more others. But really what we have to ensure is that the entire system is one that we can trust.'

Fonagy thinks it's easy for therapists to only consider the time they spend with their clients, when there is much that goes on for all the other hours when the person is not in the therapy room. He sees good therapy sessions as a potential catalyst to help someone access the social networks they have around them the rest of the time in a more positive way, but if those networks are in reality invalidating or dangerous, then that's a different problem.

'Now, what does that mean in terms of what everybody can do?' Fonagy asks himself. He is clear on the answer: 'I think we are all responsible for each other's mental health and well-being. I think the problem starts when we turn inward, and we feel that we have no social responsibilities for the well-being of others. I think mental health is inappropriately and inaccurately seen as just a subjective process, and I think making schools more mentalising, making workplaces more mentalising, making families more mentalising, making pubs more mentalising, I think this is where it's at.'

Fonagy is passionate about this. 'In our interpersonal sensitivity and recognition of others as potentially being different from us, and having thoughts and feelings that are not necessarily close to ours, recognising these as valid and to be respected is really what I think this should be about. I think we walk a very dangerous path when we abandon that kind of mutual respect, and when we wish onto others thoughts and feelings that are largely self-serving, rather than truly respectful of where they come from or what they genuinely experience.'

For Fonagy this lifts blame from families, and puts responsibility back in the community. 'Communities used to be small villages,' Fonagy says. 'Then they grew into cities. Now, because of social media, they've lost their boundaries. Now they're the entire world. And an invalidating world around one is to me a risky, dangerous thing.'

Fonagy gives some examples: 'When we can check whether the person we are talking to is telling us something that is truthful from the context they are in – their behaviour, other things that they do with others – we know we can trust them. We can open our minds and we can learn from them, about ourselves, but also about the world. When that gets taken over by a medium where assertion of opinion replaces fact, it becomes a very dangerous place.'

Fonagy's examples become more specific: 'So if I can say I think ex-President Obama wiretapped my house, and that's it, that's asserted, and I add a few times, "Believe me, believe me, believe me", you lose those facets of social communication that prior to social media used to ensure that we could trust the source of the information. So I think we are facing challenging times in terms of where we are heading, and what it is we are going to believe. I think we will probably sort it out, because human beings tend to sort things out, but it might take a little while.'

## How can we help ourselves?

Mentalisation-based treatment might not have a solution for all political dilemmas, but the approach suggests in general that what is helpful in situations where we find ourselves feeling hugely at odds with another person is not to try to interpret someone's behaviour, but simply to ask them what's going on. And in pretty much most situations as adults this can also be really useful to remember. How can I know what you are feeling if I don't ask you? How can I expect you to know what I'm feeling if I don't tell you? Even in wider political debates, how often do we truly try to understand the other political party's motivations by asking them how they've reached their conclusions? But how difficult it can be to remember this and try to do it, especially in heated situations. One of the great human dilemmas is really trying to understand ourselves and other people, but it is a dilemma that is not only worthwhile to persevere with, but that can be deeply damaging if we choose to turn away from it.

## 8

# How to train a person

Why does it make us more interested if someone blows hot and cold after a date? Why is it problematic to give in to a toddler's tantrums? Why is Facebook so addictive? How can we influence other people and ourselves to behave differently?

The answer to all of these questions lies in the science of behaviourism. The same principles that child psychologists recommend for managing a toddler can be used to understand and influence grown-ups, both others and ourselves.

## Behavioural psychology: the classics

Ivan Pavlov was a Russian physiologist born in 1849. Most of his work was on animal physiology, in particular looking at digestion and reflexes, but he is most well known for his experiments on classical conditioning involving dogs, which demonstrate a key principle of learning that applies to humans as well as animals.

Pavlov routinely rang a bell when his dogs were about to be fed, and he noticed that at the sound of the bell the dogs started to salivate, whether or not there was any food there yet. He tried this with a wide range of other noises and sensations as well, including electric shocks and the sounds of buzzers. The dogs learned, or were conditioned, to know that the bell (or whatever else Pavlov paired with the presentation of food) was associated with their meal arriving. If their food's arrival had been reliably paired with an electric shock then they would even drool when the shock was delivered independently of the food. This type of unconscious learning is the principle of classical conditioning in action, and the bell or the shock is what we call a conditioned stimulus, which is something that is responded to because of its association with something else.

Conditioning is seen in humans as well as animals. Phobias are a great example of classical conditioning, where we might learn to associate a negative reaction (disgust or fear) with a food or an animal that has previously made us sick or scared us. This conditioning can be unlearned, but it requires us to actively work at being exposed to the feared or disgusted thing to give us enough new experiences of it to help us to learn a new association.

More horrifically, homosexuality was for some time 'treated' using classical conditioning, with sexy same-sex photographs being paired with electric shocks or with medication which induced nausea.

## Operant conditioning and a man called Skinner

Classical conditioning is about repeated association leading to the linking of a thing with a reflex reaction: salivation with the bell, or fear with a photograph. In contrast, operant conditioning is much more to do with behaviour. Operant conditioning involves pairing reward or punishment with something that we do.

Burrhus Frederic Skinner, a psychologist born in 1904, explored operant conditioning using rats. He experimented with delivering a reward or punishment after specific rat behaviours, and observed that these rewards or punishments either encouraged or discouraged the rats in what they were doing. Skinner designed ingenious boxes with levers in them for the rats to press, and reinforced the rats for pressing the levers by delivering a pellet of food whenever the rat accidentally brushed against it. The rats soon learned to press the lever to receive food. Skinner used electric shocks in a similar way, but as a punishment for the rats, giving them slight shocks paired with behaviours (like lever-pressing) to discourage them.

Skinner tried all sorts of different patterns of pairing shocks or food with the rats' behaviour. Weirdly, he found that rats pressed the lever more, not when the food was delivered all the time in response to the lever being pressed, but when the food pellets were delivered some of the time, but not all of the time. Inconsistent reward was the most reinforcing thing of all.

## Reward and punishment in humans

It turns out that humans and rats are similar in their reactions to operant conditioning. We don't often think about punishment and reward except in the context of rearing children or punishing illegal acts. But in reality we are all constantly reinforcing or punishing different behaviours.

Reward can mean either giving something reinforcing, like sweets, star stickers, a glass of wine, or it can mean removing something which is disliked, like having a night off from homework, or not having to empty the rubbish bin. Punishment can mean a consequence that isn't very nice, such as time on the naughty step, or it can mean something nice being taken away, like attention being withdrawn – a parent refusing to respond to a tantrum, or a partner ignoring you when you're being annoying.

## How does it apply to children?

Unfortunately, the thing that most people remember from watching *Supernanny* or similar, is the naughty step, or time-out rule, where children are sent to sit on a step or put in a corner for a bit with no attention or interaction. Behaviour-management strategies do involve punishments like this, for a short time, but only as a very last resort, and they aren't the key feature.

Most behaviour-shaping programmes rely on reward and punishment. Punishment in a parenting context doesn't need to mean physical punishment – in fact, all the evidence overwhelmingly shows that smacking children is really bad for their self-esteem and doesn't help their behaviour to improve. The UK is behind many countries in still allowing smacking to be legal if it doesn't leave a mark, but it's really bad news. A meta-analysis in 2002[1] of eighty-eight different studies showed that physical punishment is associated with one positive effect (children immediately complying with what is being asked of them), and ten negative effects (including having a poorer relationship with their parent, having lower self-esteem, worse mental health, increased likelihood to be aggressive themselves and more anti-social behaviour). At the time, the findings were criticised as showing associations but not proving that it was the smacking that led to the negative effects, but since then more recent reviews[2] continue to find robust negative effects of smacking on children.

The same is true for harsh verbal punishment: unsurprisingly, shouting or saying mean and nasty things to children is bad for them. One study[3] followed a sample of 1,000 thirteen-to-fourteen-year-olds, interviewing the teenagers and their parents over a two-year period. The teenagers were mostly from middle-class families; social deprivation wasn't a big factor. The authors asked how much verbal discipline was used, and about the quality of the relationship between the parents and the teenagers – how much warmth the parents showed, and what the parents and children felt. They found that the thirteen-year-olds whose parents shouted or swore at them suffered more depressive symptoms in the next year than peers who were not disciplined in this way. They were also more likely to steal and fight and misbehave at school. So the way our parents use verbal discipline affects our behaviour growing up. Fewer studies have looked at the longer-term effects of this sort of punishment, but if your parent tended to shout and swear while you were growing up, it might have (understandably) made you feel sadder as a teenager than your peers who didn't have this experience.

So if punishment has at best limited use and at worst a damaging effect, then what should people be using instead? Behavioural psychology models would suggest that reward is what's left. It turns out even this isn't totally straightforward, though.

The psychologist Carol Dweck was the first to reveal that praising children *in the wrong way* can actually be bad for them.[4] It is of course much better to be praising your child than shouting at them, but the way we praise is really important. Praise can be excellent for boosting children's self-esteem, but we have to be careful about *what* we praise. Praising ability can result in children feeling constrained, feeling a pressure to get things 'right' or do things well, and feeling less able to take risks. Praising effort leads to much better effects. Dweck relates this to viewing children's intelligence and ability as something that can be grown, rather than something that is fixed and innate. Children do better when people treat them as if they have potential.

## Praising each other and ourselves

Dweck's ideas can be related to adults too. If we view our colleagues as people who can have their skills developed, rather than people who will always remain the same, we might interact with them in a different way. The same can be true of ourselves: if we make a mistake or feel we are underperforming in an area, then it is much harsher and less self-compassionate if we see ourselves as a failure in that area. If instead we view ourselves as having the potential to succeed in all sorts of different realms, but maybe needing more experience or support to really make

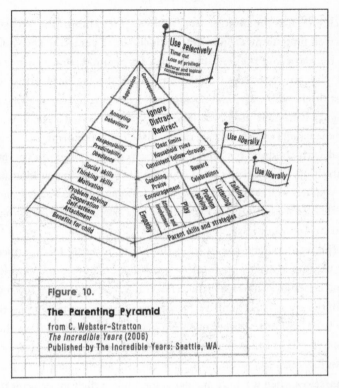

Figure 10.

**The Parenting Pyramid**

from C. Webster-Stratton
*The Incredible Years* (2006)
Published by The Incredible Years: Seattle, WA.

Figure 10. The Parenting Pyramid from C. Webster-Stratton, *The Incredible Years* (The Incredible Years, Seattle, WA, 2006)

the most of ourselves, we are going to have a much more positive experience of trying new things or taking on new roles. Praising our ability to try new things, and to persevere, might result in better results in the end.

The most successful parenting programmes, like that developed by Carolyn Webster-Stratton, emphasise spending time with your children, playing, praising (for effort), and giving them loads of attention. Consequences for bad behaviour are used, and usually involve withdrawing attention from the child, but they are used much,

much less than the positive aspects of the programme. The Webster-Stratton programme demonstrates which bits of parenting you should use the most and the least by means of a pyramid. Praise, attention and play are represented at the bottom as the big ones to use the most often, with clear limits being used where necessary; ignoring for naughty behaviour used even less, and if really needed a consequence like the naughty step or a sanction. This is really important to bear in mind with children, but also with ourselves. How often do you beat yourself up for not being good enough? How often do you internally tell yourself well done for having a go or doing well? I'd be willing to bet that for most of us this balance is vastly skewed in the direction of punishment over praise.

The word 'parent' comes from the Latin, 'to bring forth', which I like because it makes me think of growing a small person a bit like a plant, trying to encourage them to flourish. It's possible to get so caught up in the reward and punishment aspects of parenting techniques that this basic goal can be overlooked. Similarly with ourselves, if we think about trying to bring forth the aspects of ourselves we most want to encourage, this is a much more nurturing stance than just having a go at ourselves when we don't manage something in the way we want to.

## Style matters (authoritative versus authoritarian)

The principles of reward and punishment aren't the only things that matter. The style in which we use them matters too. There is a matrix widely used in research contexts to think about which style of parenting is going on. Built on work by Diana Baumring, but developed by Maccoby and Martin, the grid divides parenting styles into low and high responsiveness, and low and high demandingness. The four categories that result are set out below, and they have an effect on how we grow up. They might also make us more likely to revert to a similar style to whatever we experienced, whether in a parental role or any

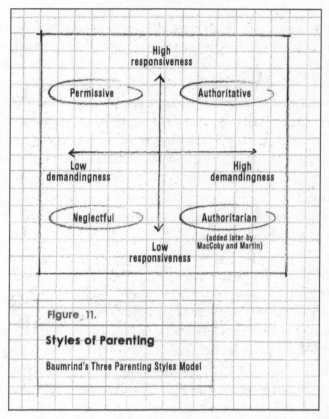

Figure 11. Styles of Parenting: Baumrind's Three Parenting Styles Model

other type of leadership role, so have a think about which one you most recognise your parents as using.

### Authoritarian

High on demandingness but low on responsiveness. An example would be a parent who expects their child to behave impeccably, but does little to respond to the child's needs when they are upset or struggling with something. Tom's memories of his childhood fit most with this description of parenting. His parents paid for him to go to a good school, bought him the latest technology to help with work as he was growing up, but expected him to get straight As and go to either Oxford or Cambridge. When he didn't get the grades they shouted at him for being lazy. They had high standards of manners which he never quite lived up to, and they didn't spend much time talking to him about his true interests. They expected him to go into law or medicine, and didn't appreciate that his true passion was actually art. He did in fact study law, and it took him several years to realise that that wasn't what he wanted to do, swapping instead to a career of graphic design. His parents struggled to understand this decision, despite him being very successful and much happier.

### Indulgent

Low on demandingness and high on responsiveness – for instance, a parent who permits their child to get away with breaking rules, and is attuned to what the child wants, maybe letting the child stay up super-late, eat in their room all the time, get away without doing the washing-up. Ingrid recalls her parents being a bit like this. She was the third child, and by the time they had had her she reckons they had just got much less anxious than they had been with her sister. She was allowed to stay up much later than her friends, and if she didn't like something she was given to eat, her mum would often make her something else. She spent a lot of her childhood running around barefoot in the garden, dressed in fancy dress when they went shopping, and danced round the aisles of stores, getting excited by the smallest things. She found it hard adjusting to rules at school, and she sometimes found the chaos at home a bit confusing, but as she grew up she was the one that most people wanted to come for sleepovers with, because her mum turned a blind eye to people drinking alcohol. She was close to her mum and dad, who were quick to spot if she was upset and to do what they could to make her feel better. She sometimes wishes that they had got her to knuckle down and work harder, and helped her stick to the rules.

### Neglectful

Low across both dimensions – low expectations but also low input. This is where, for whatever reason, parents are just not particularly involved at all, either in noticing what the child needs and responding to it, or in expecting behaviour to be a certain way. It's a bit misleading in that it isn't neglectful in the more abusive sense of the word – basic needs of food, shelter, some stimulation are still met, but more emotional needs are often shut down. Like Ingrid, Mark was allowed to stay up to whatever time he liked, but, unlike with her, his parents didn't notice when he was having a bad day. He felt pretty independent of them from a young age, and learned that if something was going badly he needed to sort that out himself.

### Authoritative

High on both dimensions – high expectations, but also high responsiveness to the child's needs. These parents expect good behaviour, but they also see when the child is upset and needs a hug, or is confused about how to navigate emotions. For Sophie, expectations at home were high: of good behaviour, good grades at school, no rule-breaking or rudeness. But she also remembers her parents bending the rules if they needed, so if she was upset, or struggling to do something, they would let her off the hook.

Much research was done on these parenting styles in the 1960s and 1970s, trying to see which style was better for children. The overwhelming majority of studies found that authoritative parenting is most helpful. It's associated with increased social responsibility in boys, increased social assertiveness in girls, increased independence (Baumrind, 1967, 1971), and increased self-esteem (Coopersmith, 1967).

But don't panic if you think you had one of the less helpful parenting experiences! Parenting style is only one factor influencing how we turn out, and even in the studies which show it made a positive effect, there is a large proportion of the positive effect that isn't explained by that factor. So Larkin's warning that our parents f*ck us up isn't necessarily true. All is not lost if you think you've had below-par parenting. It might well be worth thinking about it, though, because the examples of parenting we've experienced as children can be tempting to repeat, even unconsciously. It can be helpful to think of other examples of people who behave more in a way we would want to in terms of these styles: maybe a family friend, or a grandparent, in order to help us re-create something a bit different, either at home or at work. There's nothing magical about an authoritative style – having clear consistent boundaries and also watching out for how people feel is something we can all try to aim for in lots of different settings.

## How does it apply to adults?

As grown-ups we can all learn from the principles of behaviourism, and from what the research tells us about style, whether or not we have little ones of our own. The principles of behaviourism can be used not only in parenting situations, but also in other roles where we need to manage other people's behaviour. Shouting and incivility in the workplace is never touted as a 'best-practice' management style, but it's amazing how often people revert to it. That might be because it's what they've been used to in situations growing up, or it might just be that they don't know any other way of showing disapproval. It may simply be that their temper gets the better of them. Whatever the reasons, incivility in the workplace is bad for people's performance,[5] and downright abusive managerial styles have huge personal impact.[6]

## Reinforcement schedules for adults: drug addiction, flaky texting and Facebook

It's not just in the workplace that we can see the principles of behaviourism. There are lots of behaviours we engage in in our personal lives that we might realise are unhelpful but still find ourselves doing. Often this is because although they are unhelpful in the medium to long term, they are somehow reinforcing in the short term. How many times have you promised yourself you'll leave Facebook but then not managed to? Or how many times have you made a resolution to give up unhealthy behaviours like smoking or unhealthy eating? If you're dating, or even making new friends, how many times have you found yourself trying much harder to win the affection of someone who isn't calling or texting you back than someone who is clearly interested? Some of the processes that are at work here are the same as those that get people addicted to drugs or drink, and the same principles are important to understand if you're trying to train a toddler not to tantrum.

Let's take Facebook. I'm not convinced that it does much to improve my quality of life. If anything, it makes me more distractable, less efficient, and more likely to waste time looking at other people's lives rather than living my own fully. So why is it so hard to leave it? There might be some rational reasons why I like it: I hear about social events through it; it connects me to international friends; I enjoy seeing other people's pictures. But is there something else that hooks me in?

Remember the intermittent reinforcement schedule that Skinner discovered: when we are rewarded for doing something some, but crucially not all, of the times we do it. In Skinner's classic experiment the rats were sometimes rewarded for pressing the lever with a sip of sugar solution, and sometimes not. The reward is unpredictable. The rats totally give themselves over to pressing the lever, and don't stop easily.

People are the same. This is why it is so hard to get toddlers to do what you want them to do if some of the time you give in to their screaming fits. They learn to keep trying screaming as a tactic for getting whatever it is they want. Similarly with drugs: if you have an amazing time on a drug sometimes but not others, you are more likely to keep chasing that elusive chance of having a brilliant experience. And in romantic relationships, if someone is a bit hot and cold with you, it might make you feel bad when they are being cold, but you are still more likely to persist in trying to chase the times when they are lovely.

Facebook has a similar quality. You can comb through it for hours with nothing but mediocre photos and boring status updates about people's dinner – and then suddenly someone tags some great pictures of a fantastic holiday you went on, or there's a flurry of interesting conversation about something you care about, or lots of people like a status update of yours. Now you have lots of those little red cherries by your world, and you feel connected and approved of.

None of this is to suggest that this sort of pattern is helpful in sustaining long-term relationships with activities or people. To stay in a relationship where someone is unpredictable with their affection does not make you likely to form a healthy bond with them. And a hit-and-miss experience on social networking is not likely to lead to your days being more fulfilling if you spend more time with it. But it can be useful to understand some of the reasons why people or things hook us in.

Unlike rats, we at least have a chance of noticing when this is going on, and we are able to make more of a decision about what we do next. Unlike the rats, we can also make sure we've got our own supplies of sugar solution in reserve, by choosing to do other things we like instead.

Can you really use behaviourism on yourself? I genuinely think you can, but I'm writing this as a fully grown adult who has at times had a reward-star chart in my kitchen. For myself. I think keeping track of the small steps you make towards a larger goal is a very reinforcing way of noticing progress. I definitely go in for the carrot rather than the stick here: rewarding yourself for doing something with a little treat or luxury is a great habit to get into. I like NanoWriMo – national novel-writing month – for just this reason: it gives you a bar chart to show you how many words you write each day, and the bar chart changes colour when you hit or surpass your daily target. I used it for some of the writing of this book, and it helped me to have a run at it in a concentrated way for a month.

Equally important is giving yourself a break if you don't manage to hit your target. There is no critic harsher than the voice inside our own heads, which knows exactly which buttons to press to really make us feel bad. In the end it's totally counter-productive to spend time sitting around feeling rubbish about what we haven't done, when we could be writing a list of what to do next, identifying the key values we are working towards, or just as importantly having a bit of a rest so that we are able to carry on with the next step. Maybe even making time to feel good about the things that we have achieved: a little moment of pause to savour what we have done.

## Influencing others

Imagine if you had a behaviour-management consultant to help you manage your relationships with your partner, or your parent, or your boss. They could come in, give everyone a star chart for those behaviours that are helpful, put clear boundaries in place and remind everyone of the consequences, calm everything down, and leave you all with a peaceful situation where you felt in control. In reality, life and relationships are a lot messier than this, but in theory, behavioural principles should be helpful in shaping anyone's behaviour in a desired direction: others and our own.

We give more positive reinforcement (attention, praise) when some-one is behaving in a way we like, and we withdraw positive reinforcement when they aren't (less affection, no attention, going and spending time with someone else). We can use the same ideas from Webster-Stratton of trying to give clear positive instructions too: 'Please speak more quietly

– I don't like it when you shout at me'; 'Please do the washing-up'; 'Please say something nice about what I'm wearing or don't say anything.' It's more helpful to give clear and specific instructions concerning what we want, rather than talk about general things we don't want or like: 'You never do the washing-up'; 'You're always mean to me,' etc., although it's crazily hard to actually stick to this in the middle of an argument.

The idea of trying to spend the most time and effort on positive interactions is probably also a useful one for romantic relationships. Making time to hang out together and praising things you like in the other are pretty helpful foundations. The relationship psychologist John Gottman advocates a 5:1 rule.[7] His research suggests marriages last longer when couples balance out their communications, even in conflict, with at least five positive interactions to every negative. That can take a bit of effort if you're really enraged about someone not doing the washing-up *again*, but it's worth practising.

## Underlying values and compassion

For most of us, behavioural principles alone aren't quite enough to fully shape our behaviour. We are complex and thoughtful creatures, and making changes in our behaviours only on the basis of reward or punishment isn't as satisfying or sustainable as when we are in touch with deeper reasons for why we are doing what we're doing.

Getting in touch with the values underlying our general direction of travel in life can help us stick to behaviour changes that we want to make. Are we prioritising health? Or family? Or creativity? Or helping others? What are our guiding principles? Similarly, when we are trying to explain to others what we think should be happening, either in work or personal situations, explaining why, and trying to get in touch with the bigger values that inform our thinking, is really important, both to help people understand our motivations and get on board, and to feel authentic in what we are suggesting.

How we try to use behavioural principles, with children, with adults, with ourselves, is also really important. We can choose to use them harshly or kindly, and I'd strongly advocate the latter. Making changes is hugely difficult, and it's a process rather than a single decision-point. Just as children need help to understand and practise behaviours that help them get along with others, so we need to practise making changes over time. If we try to make a change and we 'fall off the wagon' it's no big deal: it's just human. It's all part of the process of change, and we can get back on it again. Punishing ourselves and thinking of it as a failure is really unhelpful, whereas getting back to it and seeing it as a slip-up that we can learn from is much more helpful.

## 9

# What happens when care isn't good enough?

*This chapter contains some material from an article written by the author for Mosaic Science[1]*

Not everyone's childhood is protected. Some people have really difficult experiences in their early years, and even if they leave the places and the people related to these experiences, they might have memories of them in their head as they grow up, or be caught up in unconsciously repeating familiar but unhelpful patterns.

We know from studies on attachment (see Chapter 2) that the relationships we have with our caregivers early on are enormously important. If they are good enough, then they set us up to be able to regulate our emotions, interact with others and feel brave enough to try new situations. What if the relationships we have had with early caregivers were really tricky, though? The good news is that it doesn't mean we are doomed, and studies from all over the globe agree.

## Romanian orphans

Anyone in Europe old enough to watch TV in the 1990s is likely to have a memory of the Romanian orphans. After the collapse of Nicolae Ceauşescu's time in power the world's media came in to witness the orphans left behind. Under Ceauşescu, abortion and contraception had been banned, leading to a huge surge in birth rates and a great number of babies and children who couldn't be cared for by their parents. Institutions had been created to take care of these children, but they failed to provide it adequately. Images of the children who were found in the orphanages are deeply sad: bleak rooms, packed full of small children with big eyes, pulling themselves up on their cot bars to see the Western camera operators filming them, or rocking to soothe themselves. The children had had very little individualised care: no one to hug them or comfort them, no one to sing them to sleep. Their basic physical needs were met in terms of being given food and kept warm, but their basic emotional needs for affection and comfort were not. They learned not to even bother reaching out when adults were around.

The discovery of the conditions in these orphanages prompted a rush of compassion and charity initiatives to adopt the children. The Department of Health (DoH) in the UK contacted a researcher at the Institute of Psychiatry, Psychology & Neuroscience (IoPPN), the now Professor Sir Michael Rutter.

'It's a horrible thing to have happened,' says Rutter, sitting with me in his light and airy office at the Social Developmental and Genetic Psychiatry Centre in South London. 'But, given that it did happen, one may as well learn as much as possible.

'Like everyone else, I saw the media,' explains Rutter. 'But [the research] all started because the Department of Health contacted me, to say they didn't know what was going to happen to these kids, and would it be possible to do a study, follow them through, and find out what were the policy and practice implications?'

For Rutter, this was a scientific opportunity as well as a practical one: 'This was a natural experiment.' All previous studies of children in care had had variation in the age at which the children had entered the institution, meaning variation in their behaviour and well-being might be related to things that had happened before they were in care. The Romanian orphans had all been admitted at birth or within the first two weeks of life.

Rutter's study assessed the children over time as they settled into new adoptive families. 'The findings were surprises all along the line,' he says. Prevailing wisdom at the time was that serious adversity in childhood led to a range of difficulties, not diagnosis-based, but over-arching emotional and behavioural problems. Rutter's research found something different when the Romanian children were followed up: 'There was no increase in the ordinary emotional and behavioural problems, unless they had what we came over time to call deprivation-specific patterns, meaning quasi-autism, disinhibited disorder of social regulation, often but not always accompanied by cognitive impairment or ADHD. So that was one surprise.' If the children didn't show those specific patterns of extreme difficulties with social interaction, then they didn't have problems with their emotional regulation. They were OK.

'The fact that there were no measurable sequelae[2] if institutional care stopped by six months or earlier was another surprise,' Rutter goes on. If the children were adopted out of care early enough, then they seemed to go on to develop well.

Rutter thought of this resilience in the face of adversity as a dynamic process: 'Resilience initially was talked about as if it were a trait, and it's become clear that's quite the wrong way of looking at it. It's a process, it's not a thing. You can be resilient to some things and not others,' he explains. 'And you can be resilient in some circumstances and not others.' Rutter acknowledges that 'children, or for that matter adults, who are resilient to some sorts of things are more likely to be resilient to others', but stresses that resilience is not a fixed trait. 'I tend to use the medical example of immunisation. The way to protect children against infections is either to allow natural immunity to develop, or to immunise. To protect children from all of that is the most damaging.' In other words, children need some stress in their lives, so they can learn to cope with it. 'Development involves change and challenge, and also continuity,' says Rutter. 'So to see the norm as stability is wrong. And the more one thinks about it biologically, it's got to be wrong, in that how could that be sensible?'

This suggests that there is something about the way some children adapt to and cope with adverse circumstances that enables them to be emotionally resilient. It's not the stress itself that inevitably causes problems (although in the face of enormous adversity it would be much harder to remain resilient), but it's the interaction between the stress and the ways of coping that is really important. Maybe some ways of coping are more helpful than others, and maybe some protective factors mean that the stress gets managed better.

Rutter recalls a child he saw early on from the Romanian cohort who was really struggling with his behaviour and emotional well-being, but who has now gone on to develop in a seemingly resilient way. 'He has done very well. Relationships at home are splendid. So there was a complete turnaround, and it's difficult to know precisely why that happened – but the fact that it did happen reminds you that it's a mistake to write off situations as if they can't be changed.'

## Hawaiian experimenting

Similarly hopeful conclusions have been drawn from another huge study that took place over on the other side of the globe, on the Hawaiian island of Kauai. In 1955, two researchers, Emmy Werner and Ruth Smith, started an experiment which would continue right up until the present day, spanning more than fifty years.[3, 4]

All families on Kauai who had babies that year were approached by the researchers and asked to take part. Werner and Smith, both psychologists, had become interested in which early factors set a child off on a hopeful trajectory, and which ones really get in the way of them reaching their full potential. Little did the families or the researchers know that this study would turn into one of the longest studies of child development and childhood adversity there has ever been. Nearly 700 families agreed to participate, and the researchers monitored these families and their 1955-born babies from before birth, at age one, two, ten, eighteen, thirty-two and forty. Kauai is an island pretty far from anywhere, and this meant the researchers had a fairly stable population to be following up. More recently, the research group have checked in again with the children as older adults in their sixties.

The researchers followed first the parents and then the children, finding out all sorts of things about how the cohort were doing and what sort of background they had come from. They used a mix of semi-structured interviews, questionnaires and access to community records of mental health, marriage, divorce, criminal convictions, school achievement and employment.

The researchers separated the nearly 700 children involved in the Kauai study into two groups. Approximately two-thirds were thought to be at low risk of developing any difficulties, but about one-third (201 children) were named 'high risk': born into poverty, perinatal stress, family discord (domestic violence), parental alcoholism or illness.

The researchers expected to find that those 'at risk' would do less well than the others as they grew up. In line with those expectations, they found that two-thirds of this 'high-risk' group went on to develop significant problems. But, totally unexpectedly, approximately one-third of the 'high-risk' children (seventy-two) didn't.[5] They developed into competent, confident and caring individuals, without significant problems in adult life. The study of what made these children resilient has become as least as important as, or even more important than, the study of the negative effects of a difficult childhood. Why were some of these children doing so well despite their adverse circumstances?

The study of how some of these Kauai children thrived despite early adversity is still ongoing. Dr Lali McCubbin is the current principal investigator. The daughter of Hamilton McCubbin, who worked with the original researchers, Lali McCubbin knows the history of the project well and has some Hawaiian heritage herself.

'This was a really ground-breaking study,' she tells me, speaking over Skype from her office in the US. 'What made the study unique was that despite these risk factors, that wasn't a guarantee that you would be on a certain trajectory. And in fact, what we found was there was resilience. These children were able to thrive, were able to grow, were able to develop – able to live productive and fulfilling lives.'

'A lot of these risk factors are what my father grew up with,' McCubbin adds: 'alcoholism, strict discipline, domestic violence. And I was very fortunate, I didn't grow up with that: I had a stable home, a very loving home. None of those risk factors. So I was fascinated with how you can take a risk factor inter-generationally and create not inter-generational trauma but inter-generational resilience.'

Three clusters of protective factors were more likely to be experienced by the children who did well despite being 'at risk'. Firstly, individual factors related to the child's temperament; secondly, family factors, related to having at least one figure who was consistently caring (although this could be a non-family member); and thirdly, community factors related to having a sense of belonging to a wider group.

Overall, the third of 'high-risk' children who showed resilience tended to have grown up in families of four or fewer, with two years or more between them and their siblings, few prolonged separations from their primary caregiver, and a close bond with at least one caregiver (not necessarily a parent). They tended to be described positively as infants, with adjectives such as 'active', 'cuddly' or 'alert', and they had friends at school and emotional support outside their families. Those who did better also tended to have more extracurricular activities and to get pregnant later, not in their teenage years.

The picture was complex, though, with different factors seeming to be important at different ages, McCubbin explains. At age ten, children who had had less severe birth complications were doing better than children who had had mild to moderate complications, and better outcomes were seen for children whose parents had fewer difficulties such as mental-health problems, chronic poverty or difficulties parenting. At age ten and eighteen, positive individual personality traits seemed to help, as well as the presence of positive relationships, though not necessarily with the parents. At age thirty-two and forty, marriage to a stable partner, and/or participation in the armed forces, was protective.

Strikingly, even some children who had 'gone off the rails' in their teenage years managed to turn things around and get their lives back on track by the time they were in their thirties and forties, often without the help of mental-health professionals.

'Thirty was a really pivotal point,' said McCubbin, 'Because even if you were on a trajectory such as juvenile delinquency, crime, early teenage pregnancy, having issues with drugs and alcohol, the thirty follow-up was the turnaround point. A lot of it was about finding religion, finding God and finding spirituality. Entering the military, having a sense of family, finding the right partner . . . And so not only was it about protective factors embedded in childhood, but also there was this idea that you could have a turning-point in your life.'

Many of the factors involved in the turning-point McCubbin describes, and several of the factors associated with resilience throughout the children's lives, involve relationships of some kind, whether within the context of a larger community – a school, a religion, the armed services – or in the context of one important person. 'Our relationships really are key,' says McCubbin. 'One person can make a big difference.'

Wider research suggests that there is a dose-response effect with childhood adversity, with more risk factors making children more likely to need more protective factors. But as McCubbin says, 'a lot of the research supports this idea of relationships, and the need to have a sense of someone that believes in you or someone that supports you – even in a chaotic environment, just having that one person.'

It seems blindingly obvious that how we are cared for by our parents or primary caregivers is crucial, but the growing realisation of just how important love and affection are to children has only come about in the last century. Many of the studies described in the chapter on attachment, which helped us to understand how childhood experiences can affect our adult selves, hadn't yet been published back when the Kauai cohort were born.

## Looked-after children

What if there are some children who need extra help, though – extra intervention to boost them up to the same level of development as their more resilient peers? At the moment services aim to either prevent maltreatment, or ensure that a child who is seriously affected is placed in a situation where they can be better cared for, whether this is foster care, adoption, or rarely a more institutional alternative. Generally, children only get referred for specialist help such as talking therapies or medication if they show clear signs of having developed a mental-health problem.

We know that children exposed to maltreatment are at much greater risk of mental-health problems, but we still know very little about the mechanisms involved in this risk and in resilience to it, and how these mechanisms can be targeted in treatment. Professor Eamon McCrory, Professor of Developmental Neuroscience and Psychopathology at University College London, is investigating just this.

McCrory and his team are collecting a combination of brain images, cognitive assessments, DNA and results from psychological tests, from children who have experienced a documented form of maltreatment and been allocated a social worker, and also from a control group of children. The control group have not experienced maltreatment, and are painstakingly matched with the experimental group on age, pubertal development, IQ, socio-economic status, ethnicity and sex. McCrory and his team aim to follow their cohort for two years, and longer if funding allows, to try and unpick what would predict which of the maltreated children will go on to develop difficulties, and which will be resilient.

McCrory used to work clinically for the National Society for the Prevention of Cruelty to Children (NSPCC), and he understands the clinical challenges that are involved with this population. I interviewed him back in 2015 about some of the research he is doing with his research group, and then again in 2017. 'Resources are very limited,' he explained back in 2015, 'so if you have a hundred children referred to social services who experienced maltreatment, we know that the majority of them actually won't develop a mental-health problem. But at the same time we know a minority are at significantly elevated risk. At the moment, we have no reliable way of knowing who is most at risk. So it seems sensible to try and move the focus back from the disorder to a much earlier stage in the process, to understand how a disorder emerges, as that will give us novel targets for a preventative intervention. Only longitudinal designs can give us this information.'

McCrory's research is re-framing the impact of maltreatment on the brain, previously understood by many as a form of 'damage'. It makes more sense, he argues, to understand such changes as adaptations that may help the child cope in an early adverse environment. Currently McCrory and his team are not only looking at changes in brain structures over time, but also focusing on neurocognitive mechanisms: changes in ways of thinking and processing which are seen in children who have experienced maltreatment compared to children who have not. 'Our main theoretical proposal at the moment is around a concept of latent vulnerability,' says McCrory, 'which is the idea that maltreatment leads a number of biological and neurocognitive systems to adapt to a context characterised by early stress, threat and unpredictability. Adaptations to those systems may be helpful in that context, but embed vulnerability in the longer term.' If we can identify these early markers of risk, he suggests, we may be better placed to target help at those children who need it most, before disorders have had the chance to emerge.

The team is scanning the children's brains at a range of time points to check for structural changes. They are finding robust differences in the regions of the orbitofrontal cortex and the middle temporal lobes. Despite finding these differences in this sample, they need larger-scale longitudinal studies to establish whether these changes are static, or whether they shift over time, at least in certain children. The team are especially interested in two potential neurocognitive mechanisms that are shaped by early adversity: threat processing and autobiographical memory.

Studies in both war veterans and maltreated children reveal that areas of the brain involved in processing threat, such as the amygdala, are more responsive both in the soldiers coming back from war and in children who have experienced early abuse. McCrory and his colleagues have tested children by showing them subliminally presented faces, while at the same time asking them to respond to whether a star is on the left or right of a screen. The subliminal faces are angry, happy or neutral. Maltreated children show different ways of reacting when the subliminal faces are angry. The areas in their brain involved in threat processing are more reactive, including the amygdala. It makes sense that if you have been in potential danger a lot, then your brain might have adapted to be very sensitive to threat. Calibrating brain response to threat in this way may be helpful for children (and for soldiers) in situations characterised by threat, but in the longer term such a response can be maladaptive, and increase risk of mental-health problems.

The second mechanism the team think might be important is autobiographical memory. The brain system involved in thinking about and processing memories of personal history (described in Chapter 15) might also be shaped by early traumatic experiences in a way that is adaptive in the short term but unhelpful in the longer term. McCrory's team and others have found reliable associations between autobiographical memory and childhood maltreatment. In particular, consistent with previous research, they have found that children who experience adversity tend to develop an 'over-general' style of recall, where they are less likely to recall details about their past and speak about their experiences in more general terms. McCrory and his team think that this may represent another example of a 'latent vulnerability' marker, pointing to an increased risk of future problems.

'Autobiographical memory is the process whereby you record and encode your own experiences and make sense of [them],' explains McCrory. 'We know that individuals who have depression and PTSD have an over-general autobiographical memory pattern, where they lack specificity in their recall of past experience. We also know that kids who have experienced maltreatment can show higher levels of this over-general memory pattern. And longitudinal studies have shown that a pattern of over-general memory can act as a risk factor for future disorder.

'One hypothesis is that the over-general memory limits an individual's ability to effectively assimilate and negotiate future experiences, because we draw on our past experiences to be able to predict the contingencies and likelihood of events in the future, and use that knowledge to negotiate those experiences well. So over-general memory might limit one's ability to negotiate future stressors.'

It makes sense that, if horrible things have happened to you in the past, you will want to avoid thinking about and remembering them, which might lead to a tendency to have an over-general autobiographical memory style. McCrory's team are finding reliable associations between over-general memory patterns and childhood maltreatment.

When I spoke to McCrory again in 2017 he was still working on latent vulnerability, and more evidence was gathering to support the idea. More longitudinal studies had shown associations between down-regulated reward processing and a greater likelihood to have depression later on. It seems that children who have been exposed to maltreatment are more likely to be extra-sensitive to threat cues, and less likely to notice rewarding cues, as well as less able to regulate emotion.[6] McCrory makes sense of this: 'One idea might be that if you're living in an environment that's quite unpredictable, where sources of reward are scarce or not forthcoming, it's possibly adaptive to down-regulate your expectations, because that's the reality. And you're also avoiding constant experiences of disappointment. But the reality then is that attenuated reward response is probably associated with reduced motivation, increased risk of anhedonia [a lack of happy feelings], because you're not getting the biological primer to engage in an activity and exert effort. So that's the typical response.'

McCrory thinks that with maltreatment in childhood, 'You're basically setting up a kind of cognitive groundwork for the cognitive style you see in depression, where you tend to be focusing on the negative and preferentially focusing on the negative.'

This isn't as bleak as it sounds. If we can understand the mechanisms involved more precisely, then we might be able to intervene. McCrory sees the problem as one which is likely to arise from maladaptive coping mechanisms: 'They seem to move through a trajectory where things get worse and worse. And so one idea is that they have their own ways of coping, and those ways of coping may help them in the short term, but in the long term they begin to actually exacerbate and worsen the trajectory that they're on.' Identifying the neurocognitive coping mechanisms that are actively unhelpful might allow us to teach children more helpful ways of being, or to understand what support they need in the systems around them.

'In part, the clues to that question would lie in which of the neurocognitive mechanisms seem to be most predictive of future risk?' explains McCrory. 'So if it's autobiographical memory it's likely that you'll get your preventative intervention by focusing on that. Or if it's reward processing, or if it's threat . . .' McCrory can foresee a future where it might be possible to create a detailed profile of which cognitive domains have been particularly affected by maltreatment, and tailor interventions to help accordingly.

McCrory is very clear that we shouldn't see this as a problem which is located in the child, and also that we need to be careful not to accidentally cause harm by intervening. An evidence base is crucial for ensuring that we know why we are intervening and how.

It is possible that, as well as the more specific thinking styles and patterns of brain activity, there is a more overarching common factor involved in helping individuals to recalibrate these systems in a helpful way. McCrory links this with Peter Fonagy's idea of epistemic trust – the idea that we need to be able to trust someone to teach us something about the world, and that if we don't have a trusting relationship with an individual we are unlikely to learn from them.[7] In order for a therapeutic relationship to be able to help us see the world differently, we need to be able to trust the person who is supposed to be helping us, whether that is a therapist in a formal sense or a trusted adult in another, less formal therapy setting.

'All of us update our understanding of the world in part through engaging with sources of information and people who we trust,' explains McCrory. 'And having a trusted adult may be what a child needs to recalibrate their understanding of the world, and there's lots of other evidence to support that as a key resilience factor. It's not that we're coming up with something completely new. It's that this is all coming together: it's trying to be more precise about the mechanisms that underlie the vulnerability, and thinking about how can we target those to recalibrate them, or normalize their functioning. And the framework may be a trusting secure relationship, which would fit with many things we know already.' It certainly fits with the literature on attachment, and with many therapeutic models of change.

One of McCrory's hopes is that this research will help people to begin to define resilience in a more relational context. 'I think there's a tendency to locate resilience in the child,' he says – 'somehow that it's a trait the child either has or doesn't have.'

McCrory explains resilience as more of a systemic construct: 'It's most likely a more fluid construct, which is actually an interaction between the child and adults and also peers around them. And that's what I suppose makes it such a slippery construct and hard to pin down.'

'In some way,' adds McCrory, 'resilient children have been lucky, to have had networks they have been able to engage, or networks that have been able to engage the child.'

This shifts the responsibility much more into the communities around the child. 'I think with the medical model we can easily place the pathology within the child,' said McCrory: 'the child has ADHD, or the child has depression, and that's a false sense of reality, because often those disorders are quite closely related to the social context the child is in. The same goes for resilience. It's much more likely to be a shared set of processes between the child, their parents or carers, their friends, their teachers.' Just like Fonagy in Chapter 7, McCrory emphasises the value of a positive community network around an individual.

If resilience is an interaction between the child and those people around him or her, then we all have some potential power to boost the resilience of children we know might be up against adversity. This means that we all have some responsibility, and also provides fresh hope that we might all be able to do something to help, simply by having positive relationships in our own communities with those around us.

## 10

# Thinking about thinking

Thinking is really hard to define. Bound up with ideas of consciousness and mind, every definition seems to need another one. Calling an idea to our conscious mind, rationalising, imagining, reasoning, believing, remembering, even day-dreaming . . . all of these are types of thought.

It's hard to work out exactly how and when this begins, but it is clear that our capacity for thought gets more sophisticated as we grow up. Early understanding of how children's cognitive capacities develop came from Jean Piaget in the 1930s and 1940s and, although much of it has been critiqued, the bare bones of his work still underpin many of our experimental approaches.

## Piaget's observations

Piaget worked for a while on intelligence tests for children, and while he was marking these he noticed that younger children systematically made errors in the way they answered certain questions. What interested him was not so much that they were getting things wrong, but that they were getting things wrong in a similar way. This led him to think that children's way of thinking was qualitatively different from adults'. This was a departure from previous ideas about children's cognitive ability, which people had just seen as being worse than adults'. Piaget thought our way of thinking might change as we grow up, and he went on to design a series of tasks which he got children – his own as well as other people's – to carry out. Piaget didn't set out with a specific hypothesis to test, and he didn't rigorously control the conditions, but his observations then formed the basis for a whole theory of cognitive development. In this theory Piaget saw children moving through four stages, which he catchily named sensorimotor, pre-operational, concrete operational and formal operational. It's now widely accepted that children don't develop through such clear stages, but the main concepts that Piaget identified are still really important, and lots of his experiments have been modified for use today.

## Now you see me, now you don't

If you show an infant their favourite toy, then cover it up with a blanket, what do you think they will do? Cry? Look confused? Reach for the blanket to reveal it? The answer depends on how old the child is, and on their ability to grasp object permanence – the idea that something can continue to exist even if it isn't within sight. Piaget thought that developing the concept of object permanence was a key part of what he saw as the first developmental stage, the sensorimotor stage, when infants use their sensory and motor exploration to learn about the world through trial and error. Piaget thought this stage happened in the first couple of years of life, and that this was when the concept of object permanence developed.

Piaget based this estimate on young children's reactions to their toys being covered up with blankets. He thought a lack of an ability to understand object permanence was also the reason why young children don't get as upset as older children when their parents or caregivers leave the room: because they didn't think of them when they were outside of the room as a figure that was able to come back, they didn't miss them in the same way. Subsequent researchers have shown that Piaget's timescales were off, because eye-tracking studies of babies show that they look longer at situations where objects are no longer underneath covers which have been placed on them, and then taken off. These studies suggest that infants as young as two and a half to three months old can actually understand that objects remain, even when they are out of sight.[1]

## Mountain-climbing

In one of Piaget's experiments a child is shown a three-dimensional model of three mountains, each mountain with a slightly different feature, and a load of different images of the mountains taken from different points of view. The child can walk round the mountains to start with, but then must stay still on one side. A doll is placed on the model at a certain place, and the child is asked to select the pictures which correspond to the doll's viewpoint. Young children consistently pick images that match what they themselves can see, rather than what the doll would see. As the children get older they are able to pick the pictures that look more like the doll's point of view. This egocentrism is something very typical of younger children, so when parents complain that their child is only thinking about themselves they're totally correct. When Piaget was carrying this experiment out, children four years old and below had no conception of another point of view, children of six often picked a different picture from their point of view but seldom the correct one, and seven-to-eight-year-olds managed to get it right.

Having said that, the threshold where children get better at taking other points of view in this task changes when you make the task a bit more like a story or a game. Martin Hughes did a great re-mix of this experiment in the 1970s using two police officer dolls and a child doll, asking children to hide the child doll from the police officers. This

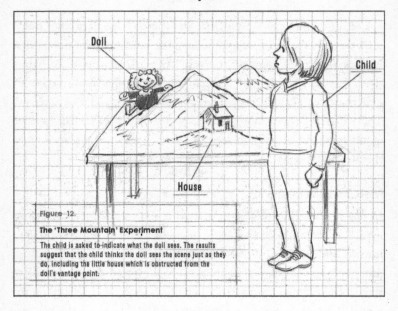

Figure 12. The 'Three Mountain' Experiment

re-working seemed to make more sense to children, and they were much better at showing that they understood that the police doll's perspective was different from their own. Most children from the age of three and a half were able to do this.

For Piaget this egocentrism was a feature of his second stage of cognitive development, which he called the pre-operational stage and put at between the ages of two and seven. He called it pre-operational because he thought children still couldn't perform mental operations like logical reasoning, although we now know that with help children can do lots more than he thought.

## Conservation issues

Take a short, fat beaker, full of liquid, and pour this into a tall, thin, beaker. The liquid fits perfectly into the second beaker even though it's

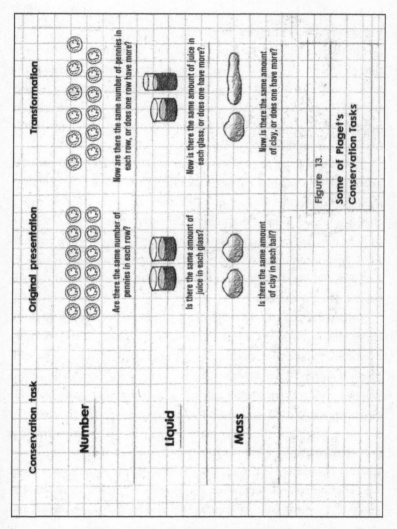

Figure 13. Some of Piaget's Conservation Tasks

a different shape. Which beaker contains more? Take two equal lines of beads, side by side, and increase the gaps in between one of the lines so it stretches further. Which line has more?

Do these seem like daft questions? The amount is the same even though the liquid is in a different-shaped container or the beads are more spread out. But ask young children this question and they reliably say that the taller glass or the spaced-out line has more. The same thing happens with a stretched-out lump of clay or a stick moved slightly forward from another.

The principle of conservation, that something stays the same in quantity even though its appearance changes, is one smaller children struggle with. It reminds me of a 1980s TV show for children with a character called Wizbit – a kind of cone-shaped magician who used to pour different-coloured liquids from one glass container to another. For young children watching these transformations of shape or size it is a bit like a magic trick, whereas as we get older we understand that the basic amount is the same.

Although Piaget thought children couldn't get the hang of this until above the age of seven, when he thought they entered the concrete-operational stage and could start doing logical manipulation with physical objects, other variations on the same experiments have shown younger children can answer questions correctly. If the line of beads described above is changed to a line of sweets, and if a 'naughty teddy' is the agent messing up the line and spreading it out, then younger children get better at answering the question, which suggests that they are better when the context is different. Over half of four-to-six-year-olds could answer questions correctly when the naughty teddy was involved.[2] The questions in the classic experiments are admittedly quite strange, and there is always a massive tendency for children to answer in a way that will please an adult, so the very fact that the child is being asked a question that presumes there is more may lead them to give a wrong answer.

## Abstract thought

The final stage Piaget thought children got to, at the age of about eleven and beyond, was that of formal operations, when he believed they had an ability to reason logically with abstract concepts and develop and test hypotheses. Again, people now think that children are able to do these things earlier, and are much less into the idea of it being a stage model, but abstract reasoning still does seem like one of the last things to develop. One example of how Piaget tested abstract thinking was a pendulum task. Children were asked to identify which factor was most important in changing the speed of the pendulum: length of string, weight of pendulum or strength of push? Older children approached this problem by testing each variable individually, whereas younger children changed all three at once. Piaget thought this capacity to think ahead and plan was evidence of abstract thinking. Any task which involves manipulating information that isn't concretely there – for example, solving logic puzzles about characters that aren't in front of the child to manipulate – can be evidence of abstract thought, whereas tasks which are more trial-and-error-based are more concrete in nature.

## Accommodation and assimilation

In Piaget's view of how our ideas about the world develop, we all have schemas that help us make sense of things. These schemas are the building blocks of bigger theories; they are the small frameworks of ideas that help us understand. I have schemas about how relationships work, about how the physical world behaves, about what a job is and how I am supposed to be at work. These schemas are based on my previous experience, but can change as a result of new experiences. The schemas affect how I view the world and influence my interactions.

Piaget thought that children create their schemas by interacting with the world around them. If they come across new information, they can handle it in one of two ways: either assimilate it into an existing schema about the world, or accommodate their existing schema to fit this new information. The same thing can occur as an adult, although we tend to have most of our schemas pretty much set up as we get older, so they are less likely to change.

## Scaffolding

If Piaget was the father of cognitive development, then Vygotsky and Bruner were the uncles. Vygotsky, a radical thinker alive during the Stalinist regime, saw learning as highly affected by the social context. He agreed with Piaget that the child was an active learner, but emphasised the child's environment much more. Vygotsky thought all children learn in what he called the zone of proximal development, or ZPD. This is the gap between what the child knows and what the child can know, with help. Bruner took Vygotsky's ideas and ran with them, introducing the concept of scaffolding, which involves a more experienced peer or adult helping the child to move through that gap in knowledge, by breaking concepts down into manageable chunks and stretching their understanding just beyond what they can do on their own. Go too far and they don't get it; don't go far enough and they don't learn anything new.

The same notions that can be important for helping children to develop can be useful for adult learning. Breaking things down into smaller, more manageable chunks; being helped by our peers to grasp something tricky; reaching for something just outside what we are currently able to do: all of these strategies are useful for any age.

## Brain and behaviour

Ever-developing methods for studying live brain activity are now resulting in increased understanding of brain development in relation to the development of thought.

Neuroconstructivism is a theory of cognitive development that argues that brain systems develop together as a consequence of the interaction of brain biology and behaviour. It goes against the idea of separate modules in the brain developing in isolation, and thinks instead in terms of a dynamic approach in which genes, brain, cognition and environment interact multi-directionally.[3] Any one of these factors can affect any of the rest, and influence how the brain develops.

This way of thinking about thinking moves away from the idea of having bits of the brain which are specialised for certain functions and separate from the rest of the brain. It means also moving away from the idea that difficulties in a specific area relate to a damaged or dysfunctional part of the brain. Instead, the brain is seen as a complex system with all parts interconnecting, and the functioning of one bit necessarily affecting the function of others.

Annette Karmiloff-Smith, who was a Professorial Research Fellow at the University of Birkbeck, wrote that 'Human intelligence is not a state (i.e. not a collection of static, built-in modules handed down by evolution and that can be intact or impaired). Rather, human intelligence is a process (i.e. the emergent property of dynamic multi-directional interactions between genes, brain, cognition, behaviour and environment).'[4] Thinking of intelligence as a process is in some ways liberating. Although it is less neat to think of intelligence in terms of brain systems rather than discrete modules, it feels easier to influence a process than to repair one part of a complex network.

## Emotional impacts on learning

Graham Music, a child psychotherapist from the Tavistock and Portman clinic in London, agrees that our capacity to develop cognitively is highly related to other areas of development, including physical and emotional development. Music sees overlaps between Vygotsky's ideas about other people helping a child to stretch to their full capability, and the importance of having a sensitive and attuned caregiver. 'For me it's very linked with secure attachment and the capacity to manage and regulate emotions,' he told me. 'Not regulate out of existence, but to manage them.'

If we haven't been helped to name our emotions, and understand that they pass, and that we can do things to influence them, then it's hard to be able to be calm enough to learn. For children who are experiencing overwhelming feelings this can really get in the way of paying attention at school, concentrating on lessons, being able to sit and listen, being able to wait patiently for break time.

The ability to defer gratification has been measured in a classic series of studies: the Stanford marshmallow experiment, carried out in the 1960s and 1970s. Walter Mischel, a professor at Stanford University at the time, devised a series of tasks for children in which they had the choice between one small immediate reward or two small rewards later, if they were able to wait for about fifteen minutes. Although the studies were called the marshmallow experiment the rewards were sometimes a cookie or a pretzel too. Following up the children many years later showed that the children who were able to wait for a larger but delayed reward did better in a range of outcomes, including school performance. Something about the ability to be patient for a delayed reward might overlap with the ability to revise well for future good exam grades, or to persevere and not give up. This experiment reminds me of my good friend Jess, who used to save her Easter eggs for literally months afterwards, having just a taste of chocolate at a time. I remember going over to her house and being mind-blown that she still had some of them left in June. Jess would have stormed the marshmallow experiment. She did also do really well in her exams, so for what it's worth there's another n=1 to add to the evidence.

## How clever are you?

It is perhaps a tiring human tendency to want to be able to measure and compare. Intelligence testing, the willy-waving of cognition, has been around for over a century. A curious example of a human construction now held up by some as an absolute truth, the idea of intelligence has been both widely criticised and staunchly clung on to.

The first person to try to create a measure of intelligence was the English aristocrat Francis Galton in the late 1800s. He thought measures of reaction times would be a good way to go, but it was a fairly crude measure and wasn't widely taken up. In the early 1900s Alfred Binet was commissioned by the French government to try to devise a way to identify schoolchildren who might need special education. His test used a range of different questions which he thought tapped into reasoning of various kinds, but involved a lot of knowledge that would need to have been taught, e.g. names of body parts, counting skills. Binet arranged the questions in order of difficulty, with the idea that children of a certain age should all be able to complete one level. The test correlated pretty well with school achievement, and it was developed and used for many years, later becoming the Stanford-Binet test when it was revised by the California-based Lewis Terman.

Many more tests were developed, some by the army to try to work out which recruits would be best at which roles, some by psychologists trying to refine the idea of intelligence by differentiating other areas of ability. Ideas of intelligence which allow it to be easily understood or categorised were as popular then as they are now, and the idea of one overall intelligence-quotient (IQ) score worked out by dividing mental age by chronological age and multiplying the total by 100 is one still used today. This fits with Charles Spearman's idea of our 'g factor' (not as sexy as it sounds). It is the concept of an overall intelligence that influences everything we do, although this idea is at odds with a lot of what we might experience: being very good at solving one type of dilemma but not at all good at working out another.

More statistical approaches to intelligence began to separate out different domains, and the most widely used intelligence tests today look at verbal ability, non-verbal ability, processing speed and working memory separately. These tests can be analysed in terms of fluid versus crystalline intelligence too, with fluid intelligence meaning reasoning ability, and crystalline referring more to knowledge that is learned.

More recently still, Howard Gardner has argued for the idea of seven different intelligences: verbal, mathematical, musical, spatial, kinaesthetic, interpersonal (social skills) and intrapersonal (self-understanding) functioning, to include the idea that we can be good at dance but bad at maths, or good at emotional intelligence but bad with words.

The current educational climate in the UK has become increasingly focused on measuring achievement and ability at a younger and younger age, with a huge impact on children's attitude to learning and levels of stress. Emphasis on testing goes quite against the literature on what promotes cognitive development. Piaget, Vygotsky and Bruner all saw the importance of the child as an active explorer in the learning process, rather than a recipient of knowledge to be tested on. Tests are stressful whether you are performing badly or performing well, and the potential pressure that can come from labels of being a 'gifted and talented' child, a new title which can be given and also taken away, is another dimension to learning that didn't exist before.

Ideas about what makes things easier for younger children to understand can be helpful for us as adults too. Remembering to break down our ideas when we explain them to others, to take them from their point of view to a place just outside their current understanding, and to allow them to learn through experience, are concepts helpful for people in all sorts of settings: business, healthcare, the arts – anything where we are trying to introduce new ideas and stretch points of view.

How about for ourselves? Is it possible to stretch the limits of our own intelligence? If we move away from thinking about points on an IQ test, and think more about different realms of intelligence, then research suggests that practice really does work. Practising different areas of ability that we think we're worse at does increase our capacity to perform in these areas, although usually what we like to practise are those things we have a natural aptitude for, so we end up getting even better at the stuff we were already good at. Research on becoming an 'expert' at a skill shows that if we put in ten years' solid and deliberate practice we are likely to be an expert at whatever it is: music, dance, writing[5] . . . This fits with evidence suggesting that one of the things that predicts school achievement is ability to delay gratification. Whatever age we are, if we can regulate our emotions and cope with the hard work of diligently repeating something over and over again in a way that is tailored to help us to get better at it, then we are likely to get good. It helps if we also get some inherent pleasure from whatever we are practising, so that the practice itself is a joy, and then it becomes more about doing the thing rather than necessarily aiming at a specific target.

# 11
# Learning right from wrong

## 'Very cunning and very wicked'

Anyone around in the 1990s is likely to remember the case of the murder of James Bulger, abducted from a Merseyside shopping centre by two ten-year-old boys, tortured and brutally murdered. The ten-year-olds were tried for the crime in an adult court, despite being only eleven at the time of trial. This resulted in huge criticism from other European countries, who thought that it violated the children's human rights, and that they should have been tried in a children's court. The British media wrote about the crime as 'evil', and the judge summed up by telling the two boys that their crime was 'unparalleled evil and barbarity . . . In my judgment, your conduct was both cunning and very wicked.'[1]

The juxtaposition of childhood innocence with extreme violence is so at odds as to be hard for us to understand, or sometimes even to want to. It becomes easier to report on an evil child carrying out an evil act than to try to understand the interweaving of factors which results in something like this happening. Yet what we know about the development of moral sense suggests that morality is a developing ability, not an innate one; one that is influenced by the environment we inhabit and, for some, perhaps also limited by their biological make-up.

## Marbles and storytelling

Experiments looking at moral development began in the 1930s, with Jean Piaget (the same prolific developmental psychologist covered in Chapter 10 on thinking). Piaget watched his children playing marbles with their friends and asked them how they understood the rules. He observed that our ability to know right from wrong, and to understand the role we can have in deciding which values to prioritise, seems to develop over time. From these conversations he concluded that between about four years old and ten years old children think that rules are unchangeable properties of the world around us, and judge the rightness or wrongness of an action by its consequences. So if Sally pushes past John and knocks him over accidentally and he breaks his leg, younger children would see this as being more wrong than older children. If Sally purposefully shoves John over but he isn't hurt at all, the younger children wouldn't think this was as serious a moral transgression, but the older ones would think it was worse. Motivation is key. At about the age of ten and over, thought Piaget, we start to understand that rules and laws are created by people, and that intentions and consequences need to be considered when we are judging the rightness or wrongness of an action.

Piaget thought that our sense of morality develops throughout childhood, partly through negotiating rules with peers. The ability to understand rules and standards of behaviour gets more sophisticated as we get older, and we become more able to judge and understand the consequences of violating moral standards. So a jury full of eight-year-olds would judge someone on what the results of their actions were, but a jury of twelve-year-olds might be asking what the intention behind those actions was.

Piaget focused on the development of moral cognition or moral thought. For him, children's growing ability to act in a moral way was related to their understanding of rules, rather than an emotional feeling. It would make sense that children's ability to reason morally got better with age, because children's ability to think in other ways also improves with age (see Chapter 10).

We now know that although Piaget's ideas about the way morality develops are worth paying attention to, moral sense develops much earlier than he thought and is much more complex, involving emotion as well as logic.

## Moral dilemmas

Some years after Piaget was watching his children play with marbles, a man called Lawrence Kohlberg got interested in moral dilemmas too, and how our view of them changes as we get older. He created a series of stories with moral twists, to see what people would prioritise. One of the most well known of his stories involved a man called Heinz (no sponsorship deal):

> Heinz's wife was near death, and her only hope was a drug that had been discovered by a pharmacist who was selling it for an exorbitant price. Heinz did not have the money. He offered what he had to the pharmacist, and when his offer was rejected, Heinz said he would pay the rest later. Still the pharmacist refused. In desperation, Heinz considered stealing the drug. Would it be wrong for him to do that?

As well as asking if it was wrong, Kohlberg asked if anything would change if Heinz didn't love his wife; what difference it would make if the person dying was a stranger; and whether the police should arrest the chemist for murder if Heinz's wife died.

Kohlberg presented his moral dilemmas to seventy-two boys from Chicago aged between ten and sixteen. He interviewed them for a couple of hours each, interested not only in their answers to the problems, but mostly in the reasoning they gave for their replies. He found that the reasons people gave for their decisions tended to change as they got older and he divided these patterns up into a stage theory of his own. Fifty-eight of the boys were also followed up at three-yearly intervals for the next twenty years.

Kohlberg's stage theory saw people moving from thoughtlessly obeying rules we are taught, in order to avoid punishment, to understanding that there might be more than one perspective on right and wrong, to internalizing whatever code we have been brought up in, through to having our own moral code based on our own values.

Kohlberg's view of humanity was a tad pessimistic, in that he thought most of us never reach the final stages of his moral framework, but instead meander along following rules that are set for us by society without thinking too much about it. His theory was based on hypothetical stories, many of which involved situations the boys would never have come across, but his stages are still interesting to consider. One feminist critique of Kohlberg's idea is that he sees morality through a justice framework, which Carol Gilligan and others argued was a very masculine way of looking at morality. Gilligan thought a more female perspective, had Kohlberg included any female participants, would have involved much more consideration of caring for others and compassion and empathy for other viewpoints. This gets us into all sorts of other issues about whether these generalisations are a helpful critique or not (see Chapter 17 for more on gender), but, whatever we think, Kohlberg did present another stage theory of morality that's worth knowing about.

## What we know now

It all seems great: we get better at moral understanding as we get older, and then we're more able to act independently in a way that means we treat each other well. Except that . . . it turns out how we behave and how we reason about moral dilemmas are only very tenuously associated. We can be brilliant at reasoning about what is right or wrong, but it doesn't mean we will behave in a moral way.[2] The moral stages that Piaget and Kohlberg came up with reflect the way we think about moral issues, but not our behavioural tendencies.[3] People also don't always make all their decisions based on one of the stages of reasoning – we might use one framework for one decision and then revert back to another stage when making a different decision.

In fact, more recent evidence suggests that both Piaget and Kohlberg were much too pessimistic in their view of what children were capable of understanding. When younger children are asked about more age-appropriate dilemmas, instead of stories about married old people and chemists, they are able to weigh up moral judgements a lot more carefully.[4] Using experiments in which the time a small child looks at something is measured (preferential looking paradigms), infants as young as three months old look for longer at a pro-social character that is being kind to others than at a control neutral character.[5] And six-month-old infants choose to interact with a pro-social character over an anti-social one.[6] Eighteen-month-old infants seem to be able to deduce intentions from other people's behaviour and try to help them – for example, helping someone to open a cupboard or pick something up that they've dropped[7] – and four-year-olds seem to be well able to distinguish between moral and conventional rules.[8]

Piaget and Kohlberg were not only pessimistic, but they also started from a much narrower set of assumptions than we have from our understanding of moral development today. Because they based their research on a logical thinking-through of moral dilemmas and rules, both theorists totally ignored the role of feelings in how we negotiate our way through moral quandaries.

## Getting emotional

The way we make moral judgements isn't just to do with how we think about a moral dilemma. If moral issues were only a logic puzzle it would be more likely we would all come to the same conclusions over emotive issues like abortion, war, euthanasia and distribution of wealth.

Experiments which manipulate our emotional reactions show that changing how we feel can alter how we judge a situation. Researchers have done all sorts of things to try to manipulate how people feel before giving them a moral dilemma to think about. Some studies use hypnotic suggestion to try to trigger a flash of disgust; some do experiments in a room with a really messy desk to try to trigger negative feeling (although this might work better with those of us who don't have a natural disposition to mess up our desks); some even spray the room with foul-smelling 'fart spray'. Volunteers in situations where they've been made to feel disgusted make significantly harsher moral judgements, compared to those in conditions without any of the disgusting cues. So emotions do seem to play a role in moral reasoning, and potentially interfere with moral judgement.

Some of the big emotions involved in making decisions about what is right or wrong are feelings of guilt and compassion. Guilt is massively intertwined with morality. It's what we feel when we recognise that we have caused pain or misfortune to someone else. Compassion is that feeling of empathising with someone else and wanting to help them; the feeling we experience when we witness someone being accidentally hurt. These 'moral emotions' seem critical in explaining differences in how people respond to moral dilemmas. The crappy way we feel after we've done something wrong, or even the anticipation of feeling that crappy, is a powerful motivator for avoiding misbehaviour.[9] Moral emotions like these have been described as '[emotions] that are linked to the interests or welfare either of society as a whole or at least of persons other than the judge or agent'.[10] They turn out to be pretty crucial in understanding why we stick to moral rules.

And here it gets even juicier. How do we develop these moral emotions? And do some of us just have more of them than others?

## Empathy and the development of moral emotions

Empathy – not just understanding, but also feeling what someone else is feeling – is, perhaps not surprisingly, linked to more pro-social ways of behaving. Children who score high on measures of empathy tend to show more comforting, altruistic and responsive behaviours towards their peers,[11] and vicarious emotional responding is also positively associated with pro-social behaviour towards peers.[12] Empathy is negatively associated with aggression, with highly empathic children being less aggressive, and lower scores on empathy being associated with more aggressive behaviour.[13, 14]

Guilt is also negatively associated with hostility and aggression,[15] and the higher our guilt-proneness the higher our empathy.[16] Guilt-proneness is inversely related to anti-social and risky behaviour, across all age ranges[18] – the more guilty we feel, the less likely we are to be anti-social or risky.

## Too much empathy?

Empathy is a really important emotion for us as a society as well as individuals. But if someone has too much empathy they might find themselves being distressed a lot of the time. Empathy might even get in the way of us being able to make clear decisions.[18] For people working in caring professions, being highly empathic and working with a number of people who are extremely distressed might lead to a phenomenon known as 'compassion fatigue', where it becomes harder to be motivated to help relieve the distress of others, and where people begin to feel a sense of burn-out. This can have huge effects on the quality of care for patients in healthcare settings, although the good news is that there are ways of reducing and preventing compassion fatigue, by boosting self-care, improving the sense of compassion in the system around the caring individuals, and remembering the key values that led to working in a caring field in the first place.

## The absence of a moral sense

While some people might be feeling too much empathy, others don't feel any at all. The idea of a psychopath is a strangely compelling one, maybe because we can project all our darkest desires into a human form – someone who really will be able to break all the moral rules and do the things we can only imagine, and worse. Or maybe because it is so chilling and so difficult to imagine transgressing all moral rules without caring. Although people are diagnosed with psychopathy, in reality, as with most clinical diagnoses, there is more of a spectrum than a clear categorisation. We all fall somewhere on the psychopathy scale, and some high-functioning psychopaths are extremely successful. In fact, the most successful psychopaths are probably the ones we struggle to spot.

Psychopathy is characterised by persistent, lifelong, severely callous and amoral behaviour. It's an interesting definition, since amorality is in itself somewhat subjective, but the upshot is it involves extreme cruelty. People who score highly on psychopathy are good at answering questions about moral dilemmas, but their behaviour is radically different, and brain scans show a different capacity to process moral emotions. Psychopaths rationally know what they should do in order not to hurt someone else, but they just don't care. They're also less good at recognizing fear and sadness in others, and they don't experience empathy in the same way most people do. It's not the same theory-of-mind deficit that we might see in Autism Spectrum Disorder (see Chapter 7), because psychopaths can understand what someone else is thinking and can work out what someone is feeling, but they don't recognise and feel emotions in the same way. It's pretty chilling stuff. Also chilling is the fact that some professions have a particularly high proportion of people who score high on psychopathic traits. Chief Executive Officers (CEOs) have a particularly high representation of psychopaths, for example.

Psychopathy seems to develop from early anti-social behaviour. Children who are diagnosed with conduct disorder (extreme naughtiness) tend to fall into two groups: those that have what are called callous and unemotional (CU) traits, who are more likely to engage in premeditated violence; and those that don't have CU traits, who are more impulsive in their aggression – responding to circumstances around them. The children with CU traits are more likely to go on to show signs of psychopathy as adults, whereas non-CU anti-social behaviour as a child can be grown out of.

Callous and unemotional traits are also more heritable, suggesting more of a genetic component. These children have unusual processing of emotions – seeing anger more easily, even if it's not there, and being less good at reading fear and sadness. Empathy is lower, although theory of mind is still there – so it's not an inability to understand what someone else is feeling, just a lack of feeling moved by it themselves.

Moral processing relies on a very distributed network of brain areas: areas that process emotions like the amygdala; areas that are involved in regulating emotions, like the orbitofrontal cortex; areas involving reward, for example the striatum; and areas involved in working out what somebody else has meant to happen – for example, the medial prefrontal cortex and temporal parietal junctions.

Brain scans of children with high callous and unemotional traits show less reactivity compared to controls when looking at others in pain, in areas related to emotional processing, especially the amygdala. Similar odd patterns of emotional processing are seen in adults with high psychopathic traits, with reduced activation in the amygdala and other related areas when they are making moral decisions. The bits that are responding differently are involved in the emotional system that is usually active during moral decision-making and action, not the cognitive capabilities that Piaget and Kohlberg were initially interested in.

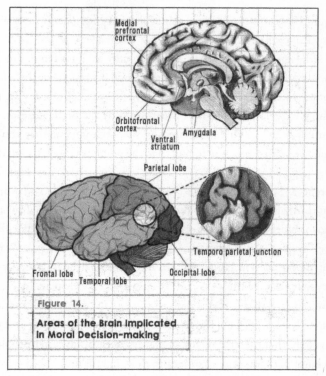

Figure 14. Areas of the Brain Implicated in Moral Decision-making

## Why psychopathy?

Professor Essi Viding from University College London has studied moral development and psychopathy for years. She initially thought she wanted to study to be a clinical psychologist, but after working as a research assistant on a study about psychopathic behaviour she got so interested that she applied for a research PhD instead. I asked her why she had become interested. 'I always find it funny when people ask,' she said. 'I always think, for goodness sake – who isn't fascinated with these really, really, emotionally utterly unusual individuals?'[19] Why do some individuals exhibit psychopathy? Viding had another straightforward answer. 'Evolution doesn't care about people being nice. It cares about propagating genes.'

Viding agrees with the idea of psychopathy as a spectrum. She uses the analogy of anxiety to explain moral sense. 'I would probably view it as your automatic feeling of whether something is right or wrong. Like anything it's going to be in part constrained by the predisposition that you inherit, and in part influenced by the environment that you get. I often use anxiety as a way to illustrate the point because it's something where people very intuitively understand that there are individual differences. So how likely you are to feel uncomfortable and worried about things hugely varies between individuals, and individuals who are prone to anxiety can use all sorts of techniques to reduce their likelihood to feel anxious – they can regulate their emotions, or they can try to apply their tension in a different way. But someone who is naturally very jittery will never be the most chilled-out person on the planet.

'Similarly, I think, if you are somebody who doesn't automatically resonate with other people's emotions and their distress, for instance, and you don't automatically feel empathy as readily as other people do, then emotional response to other people's plight is probably dampened compared to other people. And obviously it's not a fixed trend, in the sense that if you have certain environmental inputs then you can probably develop that ability up to a point – but you will never ever be the most readily resonating person on the planet. I don't think it's fixed, but I also don't think it's infinitely flexible so we can make everyone have an equally ready moral sense. I don't think that's a realistic view given what we know about every single human trait that occurs.'

## Crime and moral transgressions

Psychopathy is only one reason for crime and moral transgression. It would be nice and easy if we could say that this was the cause of all violence or deceit, but of course we're much more complex creatures than that. Crimes of passion, gradual build-ups of stress, self-rationalisation of aggressive acts . . . we probably all do things every week which are minor violations of our personal moral code. When we snap at someone on the bus, roll our eyes at someone at work, take the last of someone else's milk in the fridge and don't buy any more – we can all be anti-social to a small degree, and maybe it is just a matter of degree. Perhaps for some people the contributory factors are so great that the degree of anti-social behaviour also becomes greater, because if you've grown up in a home where you are regularly degraded or hurt, then it might be harder to understand where the lines are unless someone else shows you something different.

In the case of psychopathic traits it seems to be more of a clear difference in emotional capability. Professor Viding is clear that psychopathic individuals know they are committing crimes which hurt others, but they don't care. 'They know what's wrong, and they often say, "Well, I did it because I didn't think I would get caught, and these other people are suckers."' And they often are very aware that they are emotionally different from other people. They view other people as weak because other people are affected by these things and they don't touch them, and they really look after number one.'

## Psychopathy and what it's not

The spectrum of psychopathy is very different from the spectrum of autistic traits. Individuals who are high on psychopathic traits understand that they are hurting others but don't care. In contrast, individuals on the autistic spectrum care very much about hurting others, but their difficulties with taking other people's point of view mean that they often get things wrong.

'I think that individuals on the autism spectrum and those with psychopathy actually really nicely illustrate the importance of the moral sense or the feeling,' explains Viding. 'Individuals on the autism spectrum often really find it very hard to cope with other people's distress, provided they actually pay attention in the first place and notice that somebody was in distress, rather than concentrating on their special interest. But they don't find it easy or comfortable to be around people in distress. These are also individuals who are usually incredibly rigidly rule-bound as well.'

By contrast, explains Viding, 'Those who have psychopathic traits obviously commit way more than their fair share of crime. They're responsible for a huge percentage of crime, particularly violent crime that occurs in society, even though they are a relatively small proportion of the population. Individuals with psychopathy manipulate other people: they don't mind about other people's distress. It's not that they don't know what's right or wrong, but it's like the things that act as brakes for most of us, the stuff of "doing the wrong thing", are missing for these guys.' In essence, as described by one of Viding's students, psychopathy is a disorder of moral behaviour.

Psychopathy is also not a natural consequence of childhood maltreatment. Viding is clear about the effects on brain development of childhood maltreatment, and how it differs: 'Maltreatment affects those circuits, but it typically affects those circuits in quite a different way than what we see in individuals who are naturally predisposed to developing psychopathic traits.

'If anything, maltreatment often heightens the reactivity of those affect-processing areas. And it may be that some of the individuals who experienced maltreatment engage in aggression, but they often engage in aggression in a sort of threat-reactive rather than premeditated way.'

## Can we learn it? What helps foster moral sense?

Although the evidence now shows that even babies have some sense of pro-social behaviour, we get better as we get older at making more complicated moral decisions. By about four years old children can tell the difference between rule-breaking that hurts people and that which doesn't, but when someone accidentally hurts someone else, younger children still see this as wrong, because they don't have the theory-of-mind abilities to be able to think about intention. Similarly, as executive-functioning capabilities develop, older children and grown-ups are able to think more about conflict between intended consequences and actual consequences.

Whether it is possible to speed this up somehow, or to boost the abilities of people who have callous and unemotional traits, is another matter. With children who do have the capability to feel empathy, we can help them to understand why we consider behaviours to be right or wrong. Making it clear why someone is upset, and explaining what is OK and not OK; helping children to think about the consequences of actions and intentions for actions: these sorts of things can be helpful. Some schools programmes already use these sorts of techniques to try to reduce bullying and foster a more compassionate atmosphere.

When it comes to trying to influence children who are not feeling that it is wrong to hurt others, Professor Viding is not optimistic. 'The evidence is not incredibly good. My argument from what little we have there – what we know about fluidity of human traits – is that certainly it should be able to boost it in most children. But the degree to which we might be able to boost it in children who have high levels of these psychopathic/callous unemotional traits is an empirical question. I am a little bit sceptical as to how much you would be able to boost these individuals' empathy and moral sense, and the reason why I'm sceptical is that, if we return to the anxiety example, there's a strong intrinsic motivation for somebody with high levels of anxiety to do something about their condition, because it feels deeply uncomfortable to have that level of worry and nervousness all the time. Individuals who have high levels of CU traits do not feel bad – it's everyone around them who feels bad! So I am slightly sceptical as to what would be the intrinsic motivation of these individuals to engage in trying to feel for other people's distress.

'*We* feel it because we can't help feeling it – it happens automatically. And in many ways some of our helping behaviours are sort of motivated by empathy, but they also reduce discomfort for us. Because if we make somebody else feel better, or if we don't do the bad thing to the other person, we also don't feel horrid ourselves. And I think that if those feelings are largely absent or dampened, there is a strong motivational factor that is missing.'

## How do we help children who can't feel bad when they do wrong?

What do we do, then, to help children who can't feel bad when they do things that are wrong? Essi Viding thinks we should be trying to reward these individuals in other ways for behaving in a way that keeps society safe. 'I think to socialize these children so they behave in moral ways, it might have to happen via a different route. We may not be able to boost their intuitive moral feelings for others to a degree that makes a difference in their behaviour, but we may be able to motivate them to behave in pro-social ways because it's not going to be good for number one, and because it pays off for them in the long run. I think it's still important to get there early, because the earlier you can establish certain patterns of behaviour, clearly the more likely you are to keep somebody on the straight and narrow.'

The evidence for the effectiveness of prison for young offenders is not good, and similarly for youth offenders with psychopathic traits. Viding thinks that psychiatric provision would be better than imprisonment: still holding people responsible for their actions, but trying to influence behaviour in a more effective way. 'I think that the distinction between how you process information and how you behave is an important one,' she says. 'I am not sure how realistic or possible it is to make everyone have a certain degree of moral emotions. I think it's way more possible to have the majority of people behave in a certain societally acceptable way.

'Ultimately, as a society, that's what we ought to be interested in – that we can live peaceably among each other. I think we should care a little bit less about, "Well, did that happen because somebody is naturally altruistic and empathetic", or did it happen because they thought, "Well, actually it's better for me to do that"?'

Perhaps Professor Viding is right. If we can encourage people to avoid hurting others by using other motivations, and reduce anti-social acts and dangerous and abusive crimes, then maybe it doesn't matter whether guilt comes into the picture. It's counter-intuitive to think about, but certainly for a minority it might just be helpful.

## 12

# What do siblings do to us?

When I tell people I'm an only child I feel they do well not to flinch. There's often something: a raised eyebrow, an 'Oh', followed by a slight pause ... Only children have a bad rep – we're supposed to be spoilt and bad at sharing. We're not the only ones to suffer from a stereotype: younger children are supposed to be used to getting their own way, older ones are supposed to be serious and sensitive, and middle ones ... they're supposed to have got a raw deal all round and have a chip on their shoulder. There are positive flip sides to these stereotypes too – only children as independent, older siblings as good leaders, young children as more playful and fun, middle children as more adaptable.

The temptation to think we can predict personality from sibling status is huge. Like a weird circumstantial horoscope, we think we know a certain amount from someone's position in their family. Do we, though? How much are the stereotypes grounded in evidence?

There's no doubt that sibling relationships, or lack of them, are personally important. Many of the dramas of childhood are related to our family relationships. The experiences we were able to access were affected by who else was at home. I remember playing very happily on my own for hours, absorbed in a doll's house my dad had made me, or playing intricate games with Sophie and Emily, my two favourite dolls. I also remember games I played with the three siblings of a family who lived near me when I was young. We played Robin Hood outside, we searched for toadstools, we made a water slide in the garden. The games I played on my own were different from the ones I played with them. My friends with siblings recall their own specific memories, not always what I would predict. A younger sibling recalls playing on his own a lot because his brother was so much older, but also remembers scraps over sharing toys. Often people remember the roles they were given in the family: the clever one, the silly one – labels that can stick.

Most of us think our status in relation to siblings had an impact on the development of our personality. And sibling relationships are often the longest-lasting relationships a person will have.

## What about the evidence?

And yet ... The evidence doesn't back this up. Research on sibling effects differs. Some studies suggest there are consistent differences related to birth order, but many other studies and large-scale meta-analyses suggest there are not.

One example of a sibling study[1] from the early 1980s looked at seventy-three seven-to-eight-year-olds interacting with a sibling. The siblings were videotaped as they interacted in co-operative, neutral and competitive situations. The researchers found that first-born seven-to-eight-year-olds were more likely to be teaching their siblings and praising them, while seven-to-eight-year-olds who were the younger sibling tended to show more self-deprecation and joyful behaviour. They seemed to have more fun. Seven-to-eight-year-olds with a sibling who was closer in age to them showed more aggressive behaviour than those with a larger age gap. The authors of this study concluded that children's experience of their sibling relationship varies systematically due to structural factors such as whether they are older or younger than their siblings, and how close in age they are.

This study is a nice example of sibling studies that observe children together with their siblings. Other studies look at whether birth order has an effect that reaches outside the sibling context. At around the same time, the early 1980s, two researchers, Cecile Ernst and Jules Angst, asked 7,582 young adults to complete personality tests that measured twelve different aspects of personality, so they could see if there were any patterns in personality traits which were linked to birth order. They found no significant differences in personality traits between first-borns and second-borns in families who had two children. In families of three or more children there was only one significant difference: last-borns were significantly lower on a measure of masculinity than their older siblings.

As well as completing their own study, Ernst and Angst reviewed over thirty-five years of research on siblings. They noticed that most studies that looked for differences in personality related to birth order showed the same pattern of results they had found: no significant differences. One type of study that did tend to yield results suggesting there were differences was when the people who were answering the questions were other family members. When parents were asked to describe their children they tend to describe their first-borns as serious and responsible, and their later-borns as cheerful and independent. Ernst and Angst came up with a new idea to explain this: maybe there are sibling differences in behaviour, but maybe these differences occur around the parents.

Other large-scale studies with thousands of participants support this idea. Straightforward associations of birth order and personality traits are not found, and, if they ever are, it is often in studies where another family member responds, or where the questions asked are about the participants' traits in relation to their siblings' traits.

Judith Rich Harris, a psychologist based in New Jersey in the US, thinks the difference between some of the findings can be related to the context in which traits are studied.[2] We do develop patterns of behaviour in our families which are different depending on our relationships with siblings (or lack of), she thinks, but these patterns of behaviour do not necessarily transfer to other contexts. Just because someone is used to being the voice of authority in their sibling relationship at home doesn't mean they will automatically become a leader at school.

Rich Harris thinks this shows us where our persistent beliefs about the effects of sibling order come from: not from the research literature, but from our own lived experiences with our families. She sees these effects as being very specifically related to our own family context, though, and not something we have to take with us when we leave home. 'Birth order affects the way we behave with, and feel about, our parents and siblings', she writes: 'These behaviours and feelings are left behind, along with other mementos of our childhood years, when we leave home.'

Eileen Kennedy Moore, a clinical psychologist from New York with a particular interest in friendships, thinks the effect of having a sibling depends on the relationship you have with them.

'It depends,' she said in interview. 'It can go either way. A good older sibling can definitely help children play in a more sophisticated way – like doing imaginative play in a more interesting way than they would with kids their own age. And giving them games or rules, or even just understanding how other people react. On the other hand, there are plenty of siblings who train their siblings in bad behaviour. Yelling or hitting is not going to win you friends.'

Kennedy Moore doesn't see being an only child as necessarily being a radically different experience either. 'Many singletons have sibling-like relationships, with either cousins or close friends of the family that they're just always together with.'

## How our parents treat us

Structural factors such as birth order and age differences don't yield reliable predictions of differences, and neither do they explain why such differences should exist. Just being older or younger than a sibling in itself isn't enough to suggest an influence on character or behaviour.

Trying to pin down some of the factors that might be associated with being older or younger can help us understand a bit more why there is so much fascination with birth order, even when there is little evidence it is relevant in determining personality. Which of us isn't concerned with how fairly we are treated? The heart-rending cries of 'It's not fair!' or 'He's got more than me!' or 'It's my turn!' are familiar across the playgrounds of the world. How our parents treat us, and in particular how they treat us in relation to our siblings, might be more interesting to explore than mere birth order alone.

One group of researchers looked at just this, comparing children's temperament with maternal behaviour and family structure.[3] They found an interplay between temperament, family structure and maternal behaviour. The quality of a sibling relationship was affected by how shy or outgoing the respective siblings were, how close in age they were and also how much their mother treated them differently. More equal treatment from the mum tended to predict a less conflictual relationship. So some observed effects of being a sibling could be more down to parental treatment, and perceived positioning in the family system as a whole, rather than a direct effect of having an older or younger sibling.

## Scaffolding development

While there is no clear effect of having a sibling on later achievement and attainment, several researchers have argued that sibling relationships provide a rich source of learning, in terms of both cognitive development and emotional and social development.[4] What we are able to see our siblings do and copy from them, or even what we are able to explain to our siblings, can affect our own development. The same seems to be true for more deviant behaviours: studies of adolescent drug use suggest that having older siblings who use drugs or alcohol problematically makes it more likely that younger siblings will. The effect of having a sibling, then, is again dependent on the sort of sibling we have, more than their age in relation to ours.

## The unconscious effects of sibling relationships

The psychodynamic theorist (see Chapter 19 for more on psychodynamics) Alfred Adler, writing in the first half of the twentieth century, thought sibling interactions during childhood were really important in the specific respect of competition. He saw siblings as essentially competing for shared resources in the family, including parental affection, and thought that siblings differentiated or de-identified with each other in order to have their own individual niches, thereby reducing competition. It is interesting to apply this theory to twins in particular, since they often get referred to as 'the sporty one' or 'the clever one', even when they are identical genetically. Whether this is their family trying to differentiate them, or whether it is a recognition of the siblings themselves seeking out different roles, is hard to distinguish, but Adler would have seen this as part of the children's drive to separate themselves from each other. There is no real way of testing whether this is the case or not – it's a theory rather than an experimental finding – but it makes some intuitive sense.

Psychodynamic literature also considers sibling relationships in relation to attachment (see Chapter 2). If we have an older sibling who acts in a caregiving role to us early on, it might be that we form an attachment relationship with them. This is different from a close relationship, in that it specifically refers to that affectional tie that keeps us close to a caregiver, seeking safety from them if we feel we are in danger. There has been less work done on studying attachment relationships between siblings, which is a real shame, because there is great potential there for being able to follow these relationships up throughout life. What we might expect is that, if our sibling relationships are important to us early on, they might form the basis for some of our internal working models of relationships later on in life. Unconscious patterns of relating to others might be present in our adult selves that draw on early patterns we experienced relating with siblings. Again, this isn't as simple as being able to say that an older sibling will give you a certain internal working model of relationships: it's more nuanced than that. Neither does it mean we will inevitably treat our friends like we treated our sister, for example. But it does mean that the way we interacted with our siblings is worth bearing in mind when we try to make sense of how we relate with others today.

According to Graham Music, child psychotherapist at the Tavistock and Portman clinic in London, psychoanalysis adds a specific perspective on how siblings affect us: 'I think what psychoanalysis has helped more than anything is the ability to look at the dark sides of human nature – competitiveness, aggression, those sorts of things.' The negative side of having siblings, in his opinion, can be over-emphasised: 'I feel that there are such profound benefits that come from sibling relationships. Particularly if your sibling is born within three or four years of you, in either direction, then the likelihood is you're going to reach a whole bunch of developmental milestones earlier and better. Things like theory of mind [see Chapter 7] and those sorts of capacities, kick in earlier. The ability for ordinary reciprocity – it's been shown that having siblings has a positive effect on later peer relationships as well. And it just makes intuitive sense, doesn't it? You get embroiled in the rough-and-tumble of ordinary life and you have to give and take, and you're not the centre of the universe. You don't have to become too rigid and turn in on yourself. You're all the time a bit more other-pointing, and realising what effect you're having on other people.'

Music describes this effect less in terms of being older or younger, more as just having other people in your family that you are aware of: 'Even in a family where there's two or three siblings who aren't playing together very much, they're always in each other's mind. You always know where the other one is, and what they're up to, and when their bedtime is, and what they might be getting on their dinner table, and all those sorts of things. That is absolutely always there, I think. A lot of people actually don't feel terribly close to their siblings, and don't feel that they've got that much in common – I haven't with my siblings, in many ways – but I know that when push comes to shove and there's some kind of crisis in the family, then we'll all be there for each other.'

## Twins

Twins in particular might experience more specific effects. Being a twin is to be born into a special situation. Twins have to share resources in a more intense way than other siblings, starting from their sharing of their *in utero* environment, and going on to the demands that two same-age children make of parents being more than the demands of two different-aged siblings. One study of thirty pairs of twins used ultrasound to look at how twins behaved *in utero* and compare this with how they were after birth. Twins were seen to behave differently from each other even in the womb and the way they interacted with each other before and after birth had a continuity that could be matched when the *in utero* and postnatal footage were compared.[5] The research provided striking images of the twins jostling for space both *in utero* and in the crib, or reaching out for one another in a similar way before and after birth, which really brings home how there is a whole period of pre-birth experience that the twins have shared.

Twins often have to negotiate their childhood under more of a gaze from others around them, who might be fascinated by their twinhood. Twins also have the potential for an extremely close lifelong relationship, and a greater degree of intimacy than many other siblings. Along with this, though, as well as the task of individuating from their parents as they grow older, twins, especially identical ones, have the task of individuating from each other.

The early physical risks of being a twin are fairly widely documented, but the risks of the twin social experience on development have only more recently been considered. Studies from the first three years of life suggest that typically twins get a poorer quality of social interactions than singleton children. Young twins tend to spend more time with their twin, less time with an adult on their own, and have less sustained and more interrupted interactions with adults. They also tend to be exposed to less stimulating experiences. Parents of twins are also often having a harder time, and mothers of twins have an increased likelihood of postnatal depression.[6] All of these trends are just that – trends – and they don't apply to all families with twins, so if you are a twin, don't panic. It's helpful to realise, though, that the twin situation can compromise access to social interaction, because then we can try extra-hard to prevent these effects.

Twins have a higher risk of language delay and behavioural problems, and these might compromise how well they are able to access the pre-school experience. Some studies also suggest that twins score slightly less high on measures of cognitive ability,[7] although the difference is small. Twins often share friends, and there has not yet been enough research done to know what the effects of this might be, and how much it might be better to encourage separate friendships as opposed to shared. The lack of firm knowledge about the social effects of being a twin has led researchers quite rightly to be cautious of being too predictive about the effects of twinhood. Despite the risks of being a twin, it is clear that many twins exhibit a resilience to their situation. It might be that some twin experiences mediate these risks, or are even beneficial. More recently still, and hopefully of some consolation to twins out there, positive longer-term effects of being a twin have now also been suggested.

The idea of twins potentially gaining in social competency from their situation is relatively new. A study published in 2003 asked Finnish adolescents to rate their peers on social adaptation. Twins were rated higher than singletons. This goes against several other studies of younger children in which they were rated on their social skills by their parents and teachers. It might be that ratings by peers give a different perspective, or it might be that social skills of twins improve by adolescence to be better than their singleton friends'. Perhaps there are things about being a twin that confer social skills. Twins have to negotiate sharing from the very word go, and they constantly have another person there for social and emotional support, if they want it. Being a twin might also lead to a special understanding from a younger age that others think and feel differently: theory-of-mind skills might be better. Despite the myriad twin studies in psychological literature, relatively little attention has been paid to what it is like to be a twin, and this is one territory experiments have yet to chart properly.

## Extra siblings

Family structures today mean that we might have extra half-siblings or step-siblings that we didn't start growing up with. A lot of the research on this tends to think more about the consequences of divorce on children, rather than the effects of additional siblings. Parental divorce does tend to have a negative effect, but the effect is less when parental conflict is less, and it is almost as bad if there is high parental conflict but no divorce.[8] Studies that have been done on the effects of living in a blended family with other siblings show very small effect sizes[9] and findings are likely to be highly affected by how the situation is negotiated by the family and what the siblings are like. The same tricky emotions of rivalry and competition that are there with blood relatives are likely to come up in blended families, too, particularly if one sibling feels replaced by another, but potentially some of the same benefits can also come about, especially if the family is able to have direct and open conversations about feelings and thoughts. Easier said than done, of course, but definitely worth striving for.

## Consequences

Whatever our family situation when we were growing up, it does have an effect on us, but it isn't as clear-cut as being able to predict what older or younger children will be like when they are grown up, or what only children will be like. We might be more likely to have certain experiences if we are at a certain position in the birth order, or if we are a twin, or an only child, but we can't guarantee it. Ultimately, for each of us personally, our experiences in our family or our origin do have an effect – of course they do – but exactly what that effect is, is not predictable in the A-leads-to-B way of being able to say that if you are an older sister you will be X, Y and Z. As usual, it is more subtle and complex than that and, to my mind, more interesting.

## 13

# Playtime

## What is play?

Play is having the time and space to do whatever we want because we enjoy it. The content of play is not prescribed by anyone else. It is not assessed. It is not for anyone else, or for any other reason other than that it is enjoyable. Play can be alone, with another, or in a group. We can do it by ourselves, or with our friends. Play is fun.

Play can be all-consuming, require huge concentration and sometimes a great deal of skill. Something about it is essentially satisfying. It doesn't rely on achievement or outcome to make it worthwhile: it is good in and of itself. A word for this is 'autotelic': engaged in for its own sake. Play is autotelic. Play is enough.[1]

Despite the clarity of this definition, the study of play and of friendships has searched rigorously for their benefits, and the reasons for them. And it has seemed to find some, although the evidence base is mixed.

It's hard to devise studies that tease apart the direct impact of play, but playing does seem to be helpful to several different areas of child development. In Piaget's stage model of cognitive development, children move from sensorimotor play, where they are mostly involved in gaining mastery of their own bodies and objects around them, to more symbolic play in the pre-operational stage, to the concrete-operational stage where they start to be able to use logic and reasoning, and finally to the formal-operations stage where they can use abstract rules in games, and can reason with others in a more mature way. Similarly, Piaget observed play through the lens of moral development, and again saw a clear progression in children's ability to understand and argue about moral dilemmas (see Chapter 11), facilitated by their play with others. Vygtosky saw children playing with other children as a way of increasing their realm of understanding, and he saw peer learning as a crucial part of cognitive development (see Chapter 10). Play cuts across all these domains of learning, and it also encompasses emotional development. In playing with others we hopefully learn to recognise our own emotions and those of others, to regulate our own emotions and to respond to the emotions of our playmates. This doesn't happen quickly, and even as adults we might sometimes struggle to regulate our own emotions sufficiently to respond well enough to those around us, but well-facilitated social play can be a brilliant chance to begin to learn some of these emotional skills. Playing by ourselves can also help us learn to regulate our own moods.

Friendships help children develop cognitively, socially and emotionally. Whereas relationships with adults have an implicit hierarchy, friendships with peers of the same age are on a more equal footing. Friends are a source of fun and a buffer against stresses. Friends teach us – about the world, about how to do things, but also about how to interact with others. One study suggested that the single best childhood predictor of adapting well to adulthood is not academic achievement or behaviour in the classroom, but how adequately a child gets on with their peer group. Children who struggle to make and keep friendships are at risk.[2]

Development from playing alone through to co-operatively playing with others involves the growth of social-communication skills. Play can also be considered as a social milestone, and difficulties playing can be considered as a marker of difficulty elsewhere. Children who tend to carry on parallel playing, next to but not involved with others, and who have trouble sharing and taking turns, might be showing early signs of social-communication difficulties suggestive of autistic-spectrum conditions. This isn't to say that any child playing alone has a problem – it's easy to pathologise something that can be independent and imaginative – but just that it's worth spotting if this is the only option for a child, and if they are struggling to be able to choose to play with others for some reason.

Play is incredibly useful in its ability to scaffold interactions, and make them more pleasant and understandable for the child. Research on using puppets or games in experimental paradigms shows that children tend to perform better across a range of cognitive tasks when they are engaged in a more game-like way of doing a task, instead of feeling they are being tested on something. Play is a great way of triangulating difficult conversations as well, especially when the visual focus can be on something else, to prevent too much eye contact while talking about something potentially embarrassing. Lots of children and teenagers have really important conversations with their parents while they are in the car, where there is plenty of opportunity to be looking everywhere but directly at a parent.

Despite these likely benefits, not all studies agree that there is any additional benefit to play. Some reviews say it has no impact on children's ability to think, which is hard to imagine. Even if this is the case, play's true value might be more intrinsic, or possibly more unconscious than a direct effect on school performance or social capability.

## Cotton reels and squiggles

The theoretical approach that seems to have most to say about play is psychodynamic theory. This approach considers unconscious drives and desires, and sees the realm of play as one of the ways children can make unconscious desires and connections more explicit and acted-out. In a lecture on play, the child psychoanalyst Angela Joyce quoted a line from the Bengali poet Rabindranath Tagore: 'On the seashore of endless worlds children play.'[3] The possibilities for invention are infinite; there are no constraints.

Playing involves imagination, spontaneity, creativity and risk. It involves the mind and the body, mental and physical, and some theorists have seen it as a way for children to 'play out', and hopefully resolve, internal unconscious conflicts.

Sigmund Freud, the father of psychodynamic theory, described how his grandson played a game with a cotton reel on a string, throwing it out of sight and saying 'Gone', then reeling it back in towards him by the string and saying 'Here'. Freud's grandson did this over and over again, and Freud interpreted this play as his grandson making sense of how his mother left him every day and then returned. She was 'gone', and then 'here', too. For Freud, this everyday loss was a significantly upsetting event for the child, and he thought the cotton-reel game was a way of understanding, working through and having a sense of mastery over the temporary loss of his mother.

Sigmund Freud's daughter, Anna, was the youngest of his six children. She followed in her father's footsteps in her interest in psychoanalysis, but specialised in the analysis of children. Anna Freud devised a framework called Developmental Lines, which included six lines of development through which she thought children progressed. She acknowledged that children could also regress along these lines, and in fact would be likely to in response to trauma. One of the lines was that of play, which Anna Freud saw as moving from the body, to toys, to interpersonal relations, through to adult hobbies, and work. If we can be absorbed and creative in our play as children, then we can transfer this approach to our adult work.

Melanie Klein, of Kleinian psychoanalysis, also emphasised the importance of play. Klein thought observation of children playing could be useful at a symbolic level, to discover the child's emotional life and internal world through the games he or she plays in the external world. Klein thought that play gave a child a space to be able to act out unconscious desires, and that these could then be interpreted by the therapist and responded to in the conscious domain. Child psychoanalysis still uses play therapy and analysis of play, keeping a regular routine of time, place and toys available, and seeing what the child does with these possibilities. In a way, Kleinian child psychotherapy treats play much as adult psychotherapy uses free association with adult clients, where the person is able to say what comes to mind without filter. Both play and free association are for psychodynamic psychotherapists a way of accessing someone's unconscious.

The paediatrician and child psychotherapist Margaret Lowenfeld worked from the 1920s creating different play techniques to try to understand how children approach the world. She saw play as the child's orientation to life. She didn't think the behaviours themselves that children got involved in were as important as *how* play behaviours were engaged with: spontaneous, self-generated and self-fulfilling in and of themselves, not oriented towards reaching a specific goal. For Lowenfeld, play was more than a symbolic vehicle for thoughts and feelings. It had multiple functions of its own, related to sensory pleasure in the world and the child's own body, repetition of previous experiences, demonstration of imagined desires, and preparation for later life. Children who didn't play were missing a whole realm of pleasure, but more than this, were missing out on the functions of play itself.

Donald Woods (D. W.) Winnicott was another paediatrician and psychoanalyst working in England at around the same time as Lowenfeld. His work with children led him to develop what he called 'the Squiggle Game'. This basically involved drawing a squiggle on a piece of paper, and letting the child add to it, then interpreting what the child has added, and perhaps adding to it further. The child can then make a squiggle which the therapist adds to, and so on. Winnicott interpreted these squiggles and their additions and related them to the child's life. For Winnicott, playing was a basic form of living, and crucially important. He saw uncertainty as a threat to play: uncertainty in the environment, or in the dependability of caregivers, or even in the child's own instinctual drives. Winnicott agreed with Lowenfeld that not being able to play represented significant difficulties. Psychotherapy 'takes place in the overlap of two areas of playing', Winnicott wrote: 'that of the patient and that of the therapist. Psychotherapy has to do with two people playing together. The corollary of this is that where playing is not possible, then the work done by the therapist is directed towards bringing the patient from a state of not being able to play into a state of being able to play.'[4]

Graham Music, child psychotherapist from the Tavistock and Portman clinic in London, has noticed in his work that children who have experienced significant childhood trauma find it more difficult to feel safe enough to play, especially in symbolic ways rather than very literal games. Not being able to play might limit children's ability to work through some of the difficulties they have experienced, and might also limit their development in other ways. 'Play of course gives rise to all sorts of other developmental capacities,' says Music. 'You have to get into a developmental state where you're able to play. Without it you're in trouble, and then if you haven't got it you can't make use of all the other things like peer relationships, like symbolic play. So it's a sort of double, treble whammy.'

## Rough-and-tumble

Symbolic play isn't the only important type of play. Being able to 'rough-and-tumble' play, engaging in physical behaviours like chase and play-fighting, and feeling good about playing like this together, is also valuable. Jaak Panksepp has researched the effects of play in animals and urges us to create more opportunities for children to engage in play. In studies with rats, Panksepp found that rats with an ADHD-like profile who were given opportunities to rough-and-tumble play had reduced ADHD-like symptoms compared with rats deprived of play opportunities. Panksepp suggests that medications prescribed for ADHD actually reduce pro-social urges to engage in play, and although the drugs improve attention, there is no evidence that they promote pro-social brain development, which he thinks would in itself boost attention capacity.[5] The rise of ADHD over time does coincide with cultural shifts away from children playing outside so much and so freely as before, and, although we can't presume that this association is causal, Panksepp's research suggests that opportunities for social play are important for children's development.

## Making friends

Play can be solitary, but often it is with others, and those we choose to do our playing with can be an important part of our childhoods.

Clinical psychologist Eileen Kennedy Moore helps children and families to negotiate some of the dilemmas of childhood and writes a lot about childhood friendships. Speaking to me from her office in New York over Skype, she described why she thinks friendship is so crucial. 'You could say that friendship is the point of childhood. Friends are what make the fun times more enjoyable and the difficult times more bearable. Friendship also teaches kids important things, so whether they care about the friend, or just want to continue playing with the friend, they have to work out things like problem-solving, and conversation skills, and how to keep the game going, handling winning and losing, responding to their friends' feelings, anticipating how the friend might feel – so it really helps them to step out of their self-focus and care.'

To try to work out the effects of childhood friendships on our adult selves is tricky: it would be unethical to design an experiment where some children were denied the opportunity to make friends. One longitudinal study Kennedy Moore describes tried to measure the possible longer-term effects of having a close friend as a child. 'Fifth graders who are about ten years old who have a reciprocated best friendship – so they nominate somebody as the best friend who also nominates them as a best friend – are rated by their classmates as being more mature, more competent, less aggressive, and more socially prominent. But what's amazing,' Kennedy Moore goes on, 'is that at age twenty-three those kids with the reciprocated best friendships back when they were ten, reported doing better in college and in their family and social life, and they also reported having higher levels of self-esteem, fewer problems with the law, and less psychopathology.'

We can't jump to the conclusion that friendship earlier on causes someone to do better later on: it's likely, for example, that some of the underlying external factors that might make us better at making friends in our younger years might also lead us to having better self-esteem later – but it might also be that some of the skills we glean from good early friendships might prepare us better for our later lives.

## The skills of friendship

There are specific skills that help children to make friends, and there are common bear traps that children can fall foul of when they struggle with forming friendships. 'If you look at the development of children's friendship,' explains Kennedy Moore, 'the main ability that underlies the deepening of the friendships is perspective-taking: being able to step out of ourselves and really imagine what the other person might be thinking or feeling.' Theory-of-mind capacity (see Chapter 7) deepens as we get older, and this fits with the developmental progression we see in how children think about friendships.

'As they get older it becomes more reciprocal,' says Kennedy Moore. 'It becomes more based on genuine caring, and there's a depth to it that we won't see at the younger ages.' Whereas young children almost collect friends, or see friendships as very 'quid pro quo', later understanding of friendships is more subtle, and based on recognising that the other person may want something different. Children with social-communication difficulties such as autism spectrum conditions might find this extra-difficult, to take the other point of view. Kennedy Moore also thinks children with a diagnosis of attention deficit hyperactivity disorder (ADHD) struggle socially in particular ways, partly because a lack of attention makes it hard to pick up social cues, and partly because impulsive behaviours can be annoying to those around them but can be hard to stop.

Learning social skills can be difficult for anyone, and struggles in this domain are very common. One common tactic to try and make friends is using humour, but this can sometimes backfire. 'We all love being around people who make us laugh,' acknowledges Kennedy Moore. 'On the other hand, if you're just a little bit off in your humour, you are not funny: you are annoying. Trying unsuccessfully to be funny can really be off-putting to peers.' Kennedy Moore has some clear advice: 'A safer bet is to try to be kind. It's really hard to mess up kindness.'

Another common error children who struggle to form friendships might make is ignoring stop signals. 'We all make social mistakes sometimes, like we'll tell a joke and it's not funny, but if we stop that's OK,' explains Kennedy Moore. 'If we persist, it becomes very irritating very quickly. So one of the things that I often work on with clients is helping them to recognise those stop signals. And kids can be pretty blunt about it: "Quit it." "Cut it out." "You're being annoying." That's a stop signal, so if you're doing that you need to stop as quickly as possible, because if you persist, you're basically saying to the other person, "I don't care how you're feeling."'

How children behave in the realm of competitive games is another area Kennedy Moore thinks is important. 'The kid who cries or who has a big fuss when they're out, isn't going to get along well with their peers, and they'd rather not play with him.'

On the other hand, bragging too much about what you can do can also backfire. Kennedy Moore thinks a lot of children have a 'magnet myth' about friendship. 'They believe that they have to be so amazing and wonderful that they draw friends to them like a magnet attracts steel.' Unfortunately this doesn't work so well: 'Nobody wants to be the steel,' says Kennedy Moore. 'Those are fans, not friends, and nobody's really interested in being your fan. So really what kids need to be thinking about is not, "Can I impress everyone?" but, "How can I reach out to others in kindness?"'

What children do together also makes a difference. Sharing hobbies or after-school clubs is a way of doing fun things with others, and cultivating a common ground which can foster friendships.

## Playing alone

A lot of the literature on play describes it as starting off as a solitary activity, and, with time, developing into parallel play, where children are playing independently but near other children. Next comes co-operative play, where children engage in organised play activities that involve social roles, and for a long time playing with others in this way was seen as more sophisticated. The American psychologist G. Stanley Hall saw this evolution of play behaviours as echoing the evolution of humankind as a species, moving from solitary animals, to co-existing side by side, to co-operating with each other.

Solitary play doesn't have to be seen as inferior to play in groups, though. 'There's another kind of play that persists even at the older ages,' says Kennedy Moore, 'like building, or drawing, or creating, and I don't think we can call that immature play at all, just because it doesn't include others. There is something wonderful about being able to play on our own.' She sees benefits to solitary play that are both intrinsic and related to other useful developmental skills: 'To have a hobby that we enjoy, and to be answerable to no one but our own imagination, can be delicious. It's also great for being able to manage our moods. Because a friend isn't necessarily always available, so having that capacity to be able to entertain ourselves is healthy.'

## New playgrounds

The playgrounds children have available to them now stretch into new and digital landscapes. Where once school ended at twenty to four and that was it for the day, social media sites now mean social interactions go on 24/7. Multi-player computer gaming over the internet means playmates can be on the other side of the world.

Kennedy Moore sees social media as an extra rather than a replacement of standard face-to-face friendships. 'For the vast majority of kids, social media adds to and supports their real-life friendships. A lot of kids do feel supported by their friends on social media, but it can't take the place of real-life interaction, because it's a very attenuated form of communication. We don't have the tone of voice, we don't have the touch, so I think we want to regard it as a supplement rather than a replacement.'

We don't yet have enough research on the effects of social media to know how these types of interactions affect young people. From a clinical perspective, when things are going awry in the realm of friendships the extra social pressure of having photos and comments being uploaded into a public realm at any time of day or night is not to be underestimated. Even things that can seem relatively innocuous – a photo of some friends out and about – can be a slight if there is another friend who hasn't been invited.

As for gaming, Kennedy Moore has thought a lot about video-game culture. 'It's a new way of play that didn't exist in our generation, and for parents there are a lot of concerns.' For her the data on violent video games is not too concerning: 'The research shows that violent video games can lead to an increase in aggressive thoughts, feelings and behaviour, but it tends to be short-lived. And we don't know – having homework with long division and being asked to clean up your room can also lead to an increase in aggressive thoughts, feelings and behaviours, and we don't really know the relative difference.'

## Toys and transitional objects

The toys we choose to play with might be made specifically for the purpose of playing, or might be re-purposed household items, such as the pot that can be banged with a wooden spoon and turned into a drum. Adverts now directly target children, a powerful sector of the market, tempting them with new toys and games. Many of these are explicitly aimed at girls or at boys, and toy aisles routinely have boys' toys which include the science kits and the construction sets, and girls' toys which include dolls and princess dresses (see Chapter 17 on gender).

Grayson Perry, the artist and transvestite, certainly hasn't stuck to the gender stereotypes these toy aisles suggest. He is someone who has used the ideas of play a lot in his art, and has a separate Twitter account for his teddy bear, Alan Measles, whom he includes in some of his work. Perry, who is married to a psychoanalyst, has spoken explicitly about the links he sees between play and art. He thinks he turned to art when he realised he could no longer get the same satisfaction from play.[6] In some sense art could be seen to have similarities with play – the process of the activity in and of itself as being more important than the end result.

Grayson Perry's childhood teddy bear is often carried around by Perry's alter ego Claire, in much the same way that Linus from Charlie Brown carries round his blanket, or that we often see young children with a favourite toy or piece of soft cloth. Psychoanalysis regards this as a symbolic transitional object, helping to comfort the child as it comes to terms with the realisation that the child and its mother are separate, and that she will go away at times and he will have to live without her. The comfort provided by the cloth or toy are substitutes for the comfort provided by the mother, helping the child to manage the anxiety created by her disappearance. People joke sometimes about transitional objects, as representational of other loved ones too – almost like place markers to remind us that the person is still there in existence somewhere, even though we can't see or touch them at the moment. Again, play is seen as having a symbolic function, to help us manage bigger feelings and fears, even if we don't consciously realise it. This is an idea that is incredibly hard to test out, but one worth some consideration.

## Playing and friendships as a grown-up

Froebel called play the 'highest expression of human in the child', and I wonder if this should be extended to adults too. We usually play much less frequently than our child selves did, involving ourselves instead with the world of work, of responsibility, and of tasks that need to be done so that their goal is achieved. Even when we take up hobbies that we love to do just for their own sake, how easy it is to slip into a more goal-oriented way of approaching them: taking music exams, working up to dance performances. All great if we can still enjoy these things as ways of progressing in what we love, but sometimes it's easy to forget that we are doing things for the joy of it, not for some kind of prize at the end.

For Kennedy Moore, the value of play for both children and adults is not necessarily in any impact on other areas of life and learning. 'One recent review article said that play actually has no impact on children's ability to think or various other things,' she says. 'But I think it has value anyway, even if it doesn't improve children in measurable ways. It's a way for children to explore; it's fun; it definitely connects kids; and it's a form of creativity which I think has tremendous value, like art has value even if it doesn't improve us.'

In addition, Kennedy Moore thinks the benefits play confers on mood are substantial. 'I think it can also be a huge stress-reliever for kids. In our culture right now, or at least the north-east of the US, we're very much focused on improving children. And play lets them be them. And is an antidote to this pressure that they feel in school. It gives them a sense of autonomy too, because they get to make decisions.'

For adults too, the benefits of play could be enormous, in terms of pure enjoyment as well as stress relief, creativity, and being able to make connections between different areas of our lives in a more playful way. 'In my practice one of the most common questions that I'll ask parents is, "And what are you doing to take care of yourself?"' says Kennedy Moore. 'And with the moms in particular I usually get silence in response. Because we all feel so busy, doing so many things – but it matters. One of the most important ways we can help our children to value friendship is to show them that *we* value it: that we spend time with our friends and enjoy that company, and take care of ourselves. They're watching!'

All of the skills and tips for making friends Kennedy Moore describes can apply to adult friendships too. 'One of the things that I find so fascinating about studying friendships is, it's not like we learn all of this at age nine and then we're good for life – even as adults we're continuing to learn about friendship, in new relationships or new circumstances. It's really a lifelong issue.'

Maybe playfulness is even more of a lifelong issue from which we could all benefit. The state of mind we are in when we are playful – relaxed, open to whatever comes along, happy to join in and create and imagine: this is perhaps the antidote to most of our working mindsets, and one we could really benefit from bringing into more and more of our lives. When I was writing this book I did a six-week 'Introduction to Improvisation' course, a form of unscripted theatre which feels like pure play. All the people in the group described finding it terrifying coming along to the first session, and all described at some point feeling a joy at being silly and creative with a group of other people. I'm not saying that 'improv' is for everyone – different people like different types of play – but I found it striking not just how much fun it was, but how frightening it felt beforehand. Somewhere in our development spontaneous play often gets scared out of us, by fears of 'doing it wrong', or 'looking stupid', or 'making a fool of ourselves'. Imagine how it would feel not to have these fears, or to have these fears but mess about with different ideas and silly projects anyway – at home, but also at work. I wonder what a business or a society that truly embraced the idea of play would be like. My guess is that it could bring a huge sense of freedom, joy and creativity and might well end up with more exciting results in whatever realm it was operating in.

# 14

# School time

We spend a sizeable proportion of our lives in school, at a time when our brains are still developing and we are sponges for knowledge. No wonder it has an influence on us, academically, socially and personally. Most people still have memories from their schooldays that can bring back strong feelings: of joy, pride, warmth, and also fear, shame and sadness. When I meet up with the friends I have from this time, our conversations about what is going on now are set against a backdrop of all those other things that went on then. Old friendships bring a special pleasure, and school friendships can carry a lot packed up within them. We have a shorthand, in some ways, to understand without needing an explanation why some things would be especially hard or especially delightful. It might make us jump to conclusions too readily too, assuming we know what's going on when actually we're all quite different from our fifteen-year-old selves.

There are many different school environments: single-sex or mixed; academically selective or not; fee-paying or free; based on a specific teaching ideology or more generalist; schools aiming to help children with special educational needs or special behavioural needs; and schools aiming to coach a specific talent. From country to country the arrangements of how education works vary greatly. In the UK there is a baffling array of different school types, and whichever government is in power can have an enormous effect on what is permitted, meaning that educational provision can swing from one thing to another in a child's school lifetime.

Evidence on which schools are 'best' for children is complex, probably because the experience of going to school is affected by the personality and home context of the pupil as much as by the type of school environment. Different schools bring out different aspects of a person and have different pros and cons, and the role of the parent or carer in facilitating a child's involvement with school is also important.

Just as schools vary, so do students. As children we all have different learning needs, different physical needs, different emotional and behavioural needs and different temperaments. We come from different-sized families, we might be more or less used to having lots of children around, or more or less used to studying quietly. We come from a variety of cultural contexts, with different social rules and different economic situations, with more or less access to resources. The best-case scenario is that school can open windows of opportunity for us that we could never reach from home alone: academically, socially or emotionally.

## Relationships

Schools involve multiple relationships: between the student and the teacher, between the student and their peers, and between the student's family and the teacher. We learn most effectively as children when we have positive social interactions with teachers and with peers,[1] with high levels of mutual respect, and when teachers are trusted.[2] High-quality relationships between children and teachers and between peers are associated with students being better able to regulate their emotions, better at learning and having better social and cognitive skills.[3]

Associations persist over time. In contrast, lower-quality relationships are associated with students reporting frustration, loneliness, poorer work habits and school attendance,[4] and poorer behaviour and social competence.[5] These are all associations: we can't say for sure that better relationships cause all those better outcomes, just that they go together, but it does seem to underline the importance of trying to foster positive relationships between teachers and students.

The teacher–student relationship seems to be extra-important for some children. In some studies[6] this relationship seems to help children who have insecure attachment with their parents to do better, socially and academically, than they otherwise would. Most of the research into how children do at school and how children form better relationships with their teachers concentrates on the factors to do with the child, rather than looking at what it is that teachers can do to foster these relationships, especially with challenging children. There's some evidence to suggest that teachers being of the same ethnicity as children helps to foster good relationships,[7] but this could be for a range of other reasons. Whatever the factors are that help a child to feel understood and cared for by their teacher, it shows how important it is to allow space and time for fostering relationships in the classroom. Stressed-out, tired-out, burnt-out[8] teachers are less likely to be able to do this as well.

## Attachment theory in the child–teacher relationship

The effect of parents on children's relationship with school isn't just there in explicit interactions with teachers, or their statements to children about what school is like. The relationships we have with our parents early on give us expectations of others that we take forward into the rest of our lives. The internal working models[9] we have are applied to other relationships, including those with teachers. Associations can be found between mother–child relationships and teacher–child relationships,[10] and between a child's relationship with childcare providers like nursery workers, and their relationships with teachers later on.[11] This suggests that we might be more likely to have similar ways of interacting and similar relationships at school and at home, but what it also indicates is that, if we have a different pattern of interacting at school, and a more positive one than we've previously had, then this might be especially important.

Parents' evening is a microcosm of the relationship between school and child being triangulated by the parent. How a parent approaches talking to school about their child; whether they expect the message to be positive or negative; whether they respond encouragingly to the child or aggressively towards the school; whether they even turn up: all these things reflect the parent's attitude towards school, and affect the child's relationship with the school.

Greater parental involvement in general with school is associated with better performance academically and behaviourally.[12, 13] Volunteering to help out in the classroom, attending school events, helping with homework and openly communicating with teachers, are all factors associated with better performance at school. It makes sense that parents who have a positive outlook on school give children a more consistently positive message about it.

These associations, between better parent–teacher relationships and better child–teacher relationships, are correlations, not necessarily causation. As ever, when two variables are correlated, they might be being affected by something else: a confounding variable. Several factors affect parental involvement: parental resources; their beliefs about their own effectiveness; their beliefs about their children; their views of what a parent should do; their attitude towards the school; their experiences of their own parents and school; their cultural, ethnic, socio-economic context[14] – all sorts of things. Those parents who feel confident enough to get involved with their child's school might come from more privileged positions in other ways as well, and these things are likely to boost children's performance too.

## How does the school approach parents?

It's not just how parents and students approach school, but also how the school approaches parents and students. A large-scale study of learning in the home by Barbara Tizard and Martin Hughes looked at thirty children, from two social demographics: affluent, and less affluent, fifteen from each background. They spent time recording and observing conversations at home and at nursery, telling people they were looking at language development. In fact, they were studying how the adults were talking to the children. They transcribed and analysed the conversations, and found that conversations happening at home were a rich source of learning for all the children, both from working-class and more affluent backgrounds. In both groups the children were inquisitive and the parents were responsive. Weirdly, though, when the working-class children went to school they were more subdued, and they were in turn responded to less by the teachers. There was a gap between home and school experience which the teachers were failing to bridge, and these lost interactions were missed opportunities for learning. It's hard to say why the children from the more working-class background were more subdued at school. Maybe they had already internalised messages about what school was like; maybe they felt more nervous there for some reason: whatever the reason, it got in the way of them fully accessing what was on offer, and the teachers couldn't reach out adequately to counteract this difference.

In a similar vein, Gill Crozier and Jane Davies examined the notion of 'hard-to-reach parents'.[15] They thought this expression conjured up the idea that parents who are less involved with school may be 'difficult', 'obstructive', or 'indifferent', putting the blame firmly at the feet of parents, rather than trying to understand the barriers to engagement parents might be up against. We might do better, they suggested, to think what schools can do to be easier for parents to engage with. They re-framed the tricky issue of parents who don't participate in their children's learning as 'hard-to-reach schools'. In particular, the authors suggested that schools need to consider the different needs of different parental groups, and tailor the way they share information.

## Relationship effects on learning

One study from 2009[16] looked at children who had different attachment classifications, and how they reacted in a very specific situation where they didn't know the answer to a problem, and needed to ask for another's help.

A hundred and forty-seven children participated in the experiment when they were four years old, and then a year later when they were five. At four years old children were asked to take part in a 'novel objects' task. Child and the interviewer sat one side of a table, the child's mother and a stranger on the other. The interviewer placed a novel object on the table in front of the children – something they wouldn't have seen before, like a sprinkler or a cocktail pourer. Children were asked if they knew what the object was, and when they didn't, the experimenter asked them who they wanted to ask. The experimenter then went on to ask both their mum and the stranger for an explanation of the object. In one version of the task the mum and the stranger each gave a different made-up name for the object – for example, 'That's a snegg' – and in another version they would each perform a different mime of what they would do with the object – e.g. blowing it like a trumpet even though it wasn't a musical instrument. Children had to choose whether they believed their mum or the stranger.

In another task at age five, children were shown hybrid animal pictures. These were pictures made up out of parts of two animals – for example, 50 per cent horse and 50 per cent cow. In one version of the experiment all the animals were half and half. In another they were three-quarters one animal and a quarter another. Again, children's mums and a stranger gave answers about what the animal was called. The mums gave the least likely answer – the animal name that only had

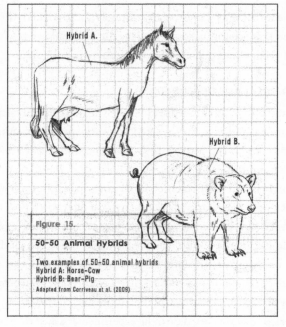

Figure 15. 50-50 Animal Hybrids

25 per cent of the characteristics. Children had to say which one they would go with.

The results showed an effect of attachment classification on the children's responses. In two of the tasks, the novel-object task and the 50-50 animal-hybrid task, where what the children could see in front of them was consistent with either adult's opinion, the children generally accepted their mum's answer over the stranger's. In the third task, the 75-25 animal task, what the children could see was more in line with what the stranger said: the stranger's animal name fitted with the majority of the animal characteristics. In this case, children generally accepted the stranger's answer over the mum's. Attachment had an impact, though: those children who were categorised as insecure-avoidant were less likely to rely on their mum's answers, and

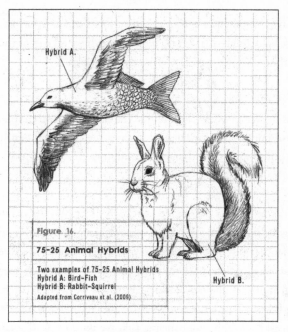

Figure 16. 75-25 Animal Hybrids

children categorised as insecure-resistant were more likely to rely on her answers. Secure children went with the mum's answer unless the evidence suggested the stranger's answer was better. They were more able to use information from both the environment and the available adults to learn about something new.

In another study,[17] babies were able to learn whether an object was something desirable or not desirable by following the cues of a stranger coming into the room – but only if that stranger first interacted with the baby. In order to develop some degree of trust in the stranger's reactions to the object, the baby first had to feel they had a social connection with that person.

If we think about teachers who were important to us, this makes huge sense. The personal relationships we have with adults who help us to learn, whether in a formal educational or an informal setting, influence how much we trust them, how much we take in of their world view, and how happy we feel to take creative risks.

## Peer relationships

School is the social world of the child. The classroom and the playground provide opportunities for friendships and for the delights of play, but also the potential for the dangers of bullying.

### Friends

Having at least one friend at school is positively associated with academic achievement, liking school and being engaged in school.[18] I still remember my best friend at primary school and the intensity of that relationship. School friendships provide a whole new arena of life learning, and the influence of the peer groups we get into can be strong. Making friends with children who are able to manage their emotions and control their behaviour means that we tend to do the same,[19] and the reverse is also true. Choices about friends really are important. It's not a level playing field, though, as children who have experienced maltreatment in their homes have more trouble than other children forming friendships. It's not surprising, if the models you have of how other people are going to treat you are frightening and unpredictable. It shows how one thing going wrong can have a spiralling effect on many other areas of our lives. It doesn't mean all is lost, but it does mean we need to work extra-hard to help children with a tough start to get onto the same playing field as everyone else.

### Bullying

Just as positive friendships get us off to a good start, and hanging out with naughty friends can get us on the wrong track, hanging out with people who are mean to us can have even worse effects. Bullying is a major stressor, whether it happens to us as children or adults. It is defined as 'repeated attacks – physical, psychological, social or verbal – by those in a position of power, which is formally or situationally defined, on those who are powerless to resist, with the intention of causing distress for their own gain or gratification'.[20] About one in ten children in the UK is estimated to experience bullying,[21] and about one in five in the US.[22]

Anyone who has ever been bullied even slightly knows how painful it can be. Being the victim of repeated bullying as a child is associated with all sorts of negative short-term and long-term consequences, including bed-wetting, difficulty sleeping, headaches and stomach aches,[23] anxiety, fear of going to school, feelings of being unsafe and unhappy at school, low self-esteem,[24] depression and suicidal thoughts.[25] There is a growing literature acknowledging that it's possible to develop post-traumatic reactions from severe bullying, and bullying can often be a trigger factor which tips people over the edge into feelings and behaviours associated with diagnoses of a mental illness. Losing lots of weight because of bullies making comments about weight or shape can tip into an eating disorder; feeling really low because of repeated meanness can tip into a depression or a tendency to self-harm. The stress of bullying alongside other stresses and other vulnerabilities can be a dangerous cocktail.

The terrain where bullying can occur has expanded with the invention of the internet. Cyber-bullying is defined as aggression that is intentionally and repeatedly carried out in an electronic context such as email, text messages, instant messages or social media websites, against a person who can't easily defend him- or herself.[26] The prevalence of this has risen hugely as digital communication becomes more and more common. Where before bullying would stop at the end of the school day, now it can continue twenty-four hours a day.

In my previous work with teenagers in psychiatric wards, under-standing what had led up to the perfect storm of their in-patient admission was often a major part of the work. We tended to use a framework that helped us to think about the long-term and short-term stressors involved, and more often than not bullying was in the mix. Hearing messages about yourself that make you feel awful about who you are, relentlessly, day after day, and experiencing school as a dangerous place to be, can just be too much to bear. The nature of bullying is that it makes people feel ashamed and to blame, which makes it harder to talk about. The fear that things will get worse if the teacher is told, or if parents find out, means that bullying can go on for a really long time, the negative messages about the person just being reinforced.

The bullies themselves deserve some sympathy, too. They don't have a good prognosis either, and they probably come from a situation which hasn't set them up so well themselves. Bullying is increasingly seen as a precursor to more aggressive behaviours later on, and bullying behaviours at school are associated with delinquency in adulthood. Leonard Eron and Rowell Huesmann[27] looked at what happened to children who were identified as bullies at age eight. Most of them had at least one criminal record in adulthood. Similarly, Eron and colleagues[28] found that 25 per cent of children identified as bullies in their early school career had a criminal record by the time they were thirty, and Dan Olweus found similar associations between bullying at school and arrests later on.[29] Children who bully at school are more likely to report having poorer physical health and increased risk of mental-health problems in adulthood.[30, 31] This might be linked to the reasons behind someone bullying. Jeanette Winterson, in her book *Why Be Happy When You Could Be Normal?*, writes about the reasons why she bullied other children when she was younger, and they boil down to a deep unhappiness: 'I realise my behaviour wasn't ideal, but my mother believed I was a demon possessed and the headmistress was in mourning for Scotland. It was hard to be normal.'[32]

Even less often spoken about is the possibility that teachers can be involved in bullying, either as the victim of targeted bullying by schoolchildren, or as the perpetrator of bullying in the classroom. There is much less research on this area, so we don't know enough about possible causes and effects of bullying involving teachers in either direction.

## Girls and boys and bullying

There has been a good deal of study in the UK, US and Europe, exploring which types of bullying girls and boys are most likely to be involved in, either as victim or perpetrator. The literature suggests that sex makes a difference to the sort of bullying behaviours. Girls are more likely to engage in and to be subject to indirect bullying. This includes verbal aggression, gossip, and social isolation. Boys are more likely to engage in and be subject to more direct forms of bullying. This includes physical aggression and destruction of belongings. This only speaks to a general trend – girls can also be subject to physical bullying and boys to verbal aggression – but it is worth being aware of, partly because the effects of the two types of bullying seem to be slightly different.

One study that tried to tease apart the effects of these different types of bullying on the different sexes used data from over 10,000 students in the US.[33] While both types of bullying are really upsetting, the researchers found that girls who are bullied by either direct or indirect methods have lower self-esteem in relation to their academic achievement than boys subject to these types of bullying. Direct bullying didn't statistically affect boys' self-esteem about work, but did seem to impact negatively on their academic achievement. Indirect bullying affected both boys' and girls' self-esteem about work, but had a much bigger effect on girls. Girls seem to be especially vulnerable to this effect of bullying. The authors didn't have the data to look at longer-term effects of the different types of bullying, but bullying of any sort is really bad for the victim's self-esteem, sense of self-worth and sense of being able to do anything about the problem.

## Effects of school experiences

The obvious consequences of not doing so well at school academically are that we have fewer qualifications and might find it harder to find meaningful paid employment that we enjoy. Further-reaching than this is the erosion of self-esteem that can result from an experience of a learning environment we find confusing or punitive. This can be so great that it removes the desire to even try to succeed. The consequences of feeling that we don't fit in socially at school, or of more pernicious bullying, can also be long-lasting, affecting how we approach making friendships in adult life. Some of the most powerful effects can be of the unconscious beliefs about ourselves, the world around us and other people, that we develop through our experiences of school. One of these is the effect school can have on our own attitudes to learning. This persists in its importance as we grow older.

Whether we approach learning a new skill with delight or trepidation affects our experience of, and probably also our persistence with, learning. Learning something new is a risky business. Just think of going to a new dance class as an adult, and all the anxieties that can bring up. Learning environments can do much to relax us or to make us more tense, but this will differ from person to person. The libraries of Oxbridge are amazing places to learn in and concentrate, but they can also be paralysing in their grandeur if they aren't what we're used to. Just as we have internal working models of relationships from our families of origin, so we have internal working models of learning environments and cultures from the schools that we went to. Think of where you went to school, the experiences you had: how did those affect you?

Michael Rosen, the children's writer and former Children's Laureate of the UK, has commented extensively on how the school environment can influence children's learning. I asked him what he thought made a good enough learning environment. 'A good enough learning environment is one in which children will learn!' was his reply. 'Tautological, I know, but we have invented quite a lot of aspects of environment which discourage some children from learning. We persist in immersing children in large groups in close-packed environments with not a lot of physical exercise, not much choice and with very little autonomy. Some children cope with this; many don't.'

When we think of the systems we create around children to organise their learning and measure its effectiveness, it is interesting to think how much of their design is created for the benefit of children's learning, and how much to make it easier for us to manage and assess children. In the UK the amount of mandatory assessment has increased, and the age at which children sit exams has got younger. Rosen sees problems with the exam culture: 'The point is, that the more testing and exams we create, the more we "prove" that some children are not "good enough", no matter what.'

Graham Music from the Tavistock and Portman clinic in London, whom we met in previous chapters, has a similar view on exam culture: 'Tests . . . The only people they help regulate is the politicians. I can't see how that helps at all.' Music sees an effective school environment as encompassing emotional and physical well-being, referencing, for example, experiments which have shown positive effects from physical exercise on academic achievement.[34]

He sees negative effects from the pressure children and teachers are put under by constant assessment. 'Emotionally understanding the kids is going out the window. There's so much pressure to get through the curriculum, and then that stops you actually being attuned to where the kids are. So actually it's moving away from Piaget' (see Chapter 10), 'away from theory: "I've got this template that you've got to fit into, and if you don't it means that either I'm a bad teacher, or you're a bad pupil." And it's just hopeless, really.'

For Music it comes back to relationships, and school provides another opportunity for a child to form positive relationships: 'I think the most important thing for a child's development is to be believed in, and feel that they're really thought about in a caring, loving, compassionate state of somebody else's mind. So that they can then internalize that and take it with them. So much of what kids are receiving these days is pressure, pressure, pressure. For a kid to flourish, you want them to be happy, at ease, confident, not overly narcissistic or feeling they're the centre of the universe. They don't have to be omnipotent, or think they know it all or have to get it all right. I think that comes from having adults around you, obviously mainly your parents, but your teachers as well – that you know they've got your interests at heart somehow.'

The extent to which our early schooling sets us off on a life trajectory can be scary to think about – a bit *Brave New World* in its determinism. The proportion of the British government who went to public school, for example, is disproportionately large. There will, of course, be other factors at play – for example, level of affluence, and family expectations – but the experiences of large public schools might also be familiarising people from a young age with the debating skills and environments they might need or encounter in government. If we are made to feel at an early age that we have the right to be comfortable in a variety of environments, we grow up feeling that we belong, and can take up our own space in any setting we choose. Similarly, if we have a sense early on of a permission to be curious and to ask questions, then we are more likely to be able to fully engage with the curiosity needed for full understanding, as opposed to a surface-level rote-learning of answers.

As with all earlier experiences, nothing gets set in stone for us as a result of what happens early on, but it might take more conscious effort and sometimes a bit of help to break away from the most likely trajectory. It is always possible to think about our childhood, and make sense of early experiences in a way that allows us to have a wider range of choice as grown-ups: to be able to respond thoughtfully, instead of reacting without thinking at all. If there are areas where we are lacking in skills, we can make the effort to learn them. Things that might not come naturally can nonetheless still be taken on board.

Ultimately, whatever school we go to, if we can come out with a love of learning and some social skills we're likely to do OK. The philosopher Alan Watts talked a lot about learning as being a process, rather than a race towards an end result. Unrelenting examinations can lead students to be focused on the grades they achieve, rather than the things they learn about. Watts compared this to realising that the point of a musical piece is not to get to the end fast enough, but to savour the whole composition: 'We thought of life by analogy with a journey, with a pilgrimage which had a serious purpose at the end . . . but we missed the point the whole way along: it was a musical thing, and you were supposed to sing or dance.'[35] Perhaps the key for all of us is trying to remember that, and trying to instil an appreciation for loving what we do, and doing what we love along the way as early on as possible – enjoying the learning in and of itself, instead of getting caught up in the rush for exam grades or other badges of life achievement.

# 15

# Remember, remember

## What is memory?

Before we consider how our memory develops, and at the risk of disappearing down a philosophical rabbit hole, it's worth stopping to consider what we mean by memory in the first place. There are multiple definitions, and a general consensus is that these reflect different memory systems in the brain for different memory functions.

Robyn Fivush, Professor of Psychology at the University of Emory in Atlanta, USA, has studied memory for many years. She agreed to be interviewed about how she sees memory and its development. 'I think the lay public, when you say memory, think autobiographical memory: memory of my personal experiences. But in the literature, memory and knowledge are not that clearly distinguished. Anything we know, we had to have learned at some point, either by hearing it, or by experiencing it, so that kind of information is often called semantic memory, which is just all the knowledge we have about the world.'

Semantic memory, such as, 'Paris is the capital of France', and, 'Shakespeare wrote *Romeo and Juliet*', doesn't tend to be the sort of memory we think about when we say 'I remember . . .' But this sort of knowledge is often used in research studies about memory.

Similarly, memory of everyday series of actions like those involved in driving a car, or riding a bike, we tend to descibe as knowledge rather than acknowledging it as a type of non-declarative, procedural memory. I don't say, 'I remember how to ride a bike.' I say, 'I know how to ride a bike.'

There are also distinctions in the literature between short-term, long-term and working memory. Short-term memory is the capacity to hold information for a short time, but not manipulate it in any way. Working memory is also a short-term type of memory, but this system can hold information for processing, so it can be manipulated – for example, holding a maths question in your head while you calculate the answer using mental arithmetic. Long-term memory is the ability to hold information for longer periods of time (episodic, semantic and procedural memory are all seen as categories of long-term memory).

## How do we measure memory in children?

Studying memory capabilities in infants has often used habituation: the gradual reduction in the strength of a response as the result of repetitive exposure to something. An example would be showing a baby a picture of a green apple over and over again, and then presenting it with a picture of a red apple and seeing if it looks for longer or shows any signs of surprise. This is useful for scientists trying to work out what infants who can't speak might be able to remember. Infants older than about eight to ten weeks old tend to look at new things for longer than things they have seen before, suggesting they can remember. Infants are repeatedly exposed to a stimulus until they look at it a certain amount less, then shown something new. Recognition of the old object or picture is usually inferred if the percentage of time looking at a new stimulus compared to the old exceeds chance (50 per cent).

This type of experiment has to be done really carefully to make sure that the old and new stimuli are similar in terms of complexity and dynamism. Children prefer to look at more complex things and at more dynamic things. Another criticism of this technique is that preferential looking tends to increase for stimuli that are slightly familiar, and then decrease when stimuli are fully familiar. It could be that with a slightly familiar object babies are comforted, or still interested and not bored. Whatever the reason, it makes these preferential-looking habituation tasks a bit harder to interpret straightforwardly.[1] Similarly, there might be ways that stimuli differ that are hard for us to see as adults, but which do make a difference to children. It's not as simple as just asking someone if they remember something or not.

Babies who have heard their mothers' voices a lot, or particular pieces of music, while they were in the womb, seem to remember these sounds for a short time after they are born. Fivush calls this an implicit type of memory – a recognition. This recognition or familiarity is different from what Fivush thinks of as more conscious types of memory and recall. 'I don't go, "Oh, yeah, I remember hanging out in mom's womb and hearing this, you know,"' she explained. It is important, she thinks, that we realise this is a more primitive type of memory. 'As adults we tend to think, "If I remember this way then that's the subjective conscious experience for infants," but I don't think it is. I think that that's a very primitive form of memory, just like our pets remember us, but our puppies don't say, "Oh, there's Lucy, she left this morning at 9 o'clock, and now she's home and I'm so glad" – but your dog remembers you in some sense you know. So I think we use the term to cover a lot of different territory.'

Much of the research done with young-infant memory performance also considers recognition more than conscious recall. Nonetheless, there are correlations with memory performance later in childhood, suggesting an overlap of memory systems.[2] One experiment with over 200 infants (59 pre-term infants and 144 controls) saw the children three times in their first year of life, and then again at age twelve, twenty-four and thirty-six months. Pre-term and full-term infants were matched on birth order, ethnicity, parental education and socio-economic status, and three types of memory were assessed: immediate recognition, using a visual paired comparison task (habituation); recall memory, by getting infants to imitate a sequence of actions; and short-term memory, using a habituation experiment that had increasing numbers of items in it as the infants got older.

Just like adults, infants and toddlers performed differently on different types of memory task. Longer-term and shorter-term tasks were dependent on different underlying factors, and recognition was linked to two separate processes: recollection and familiarity. Babies who were premature did worse in the memory tasks, especially on recollection. This fits with what we know about the hippocampus being less well functioning in premature babies, since the hippocampus is a brain structure involved in memory. As infants got older they were able to remember more things in their working memory.

'That's one of the things that changes developmentally,' explains Fivush. 'We have what's called working memory: what we can hold in mind at any given time. That actually increases across early childhood, just because the brain develops, and because we become better able to regulate our cognitive functions. So we actually can hold more information at a time. We can also organize it better, which is one of the reasons that memory has become more stable and enduring. But that really happens, I would say, through the first ten years of childhood.'

## Childhood amnesia

Childhood amnesia sounds terrifying, but in reality we all have it. What was your first memory? Let your mind go right back ... What comes into your head? Was it something you were scared of? Or that you really liked? Something out of the ordinary? Roughly how old would you say you were?

It's hard to pin down early memories. It's a bit like trying to catch something that is continually moving. Early memories can be slippery and elusive. Often we think we remember stories that have been told to us again and again, or scenes from familiar photographs. Often early memories have some notable degree of emotion attached to them. I remember being scared of a daddy-long-legs in the conservatory of the house we lived in when I was about three or four, and a flood in the same house – me pretending to swim as I walked about in the water in my welly boots with my skirt tucked up in my knickers.

I can't remember much before this age. Sadly, I can't remember any of the American road trip my parents went on while I was little: I can't remember learning to walk in Texas, or going to Disneyland, or visiting the forest of sequoia trees with enormous trunks.

I'm not unusual in not being able to remember anything before the age of about three. This phenomenon is called infantile amnesia: the forgetting of most experiences we have before the age of three or four. If you ask children what they remember, though, they tend to be able to go further back than adults. Some studies suggest that children have 50 per cent accuracy in relation to events that occurred before the age of two, but most adults remember nothing of that period.[3] Up to about ten years old children can remember lots of what happened before. Memories appear to fade as children approach adolescence. By age eleven, adult levels of amnesia are already evident. What on earth happens?

This is a mystery that has not been solved. One possibility is that, as the frontal cortex undergoes massive reorganisation in adolescence, the earlier memories are somehow lost. Another possibility is that, because the frontal cortex was so under-developed at the age of the memories happening, those early memories weren't properly laid down. Since it is the pre-verbal memories that we tend not to recall, perhaps something about the lack of verbal representation of our experiences means these memories go. What we do know about the brain and memory is that it involves multiple brain areas, and many of these are continuing to develop for some time after we are born. The hippocampus continues to develop until age four, and frontal lobes continue to proliferate and then prune neurones and white matter up to our late teens and early twenties.

Fivush and her colleague Catherine Nelson have argued that it is less helpful to think of childhood amnesia as a point before which you remember and after which you can't, preferring instead to think about a gradual emergence of memory through childhood. 'You know most people have very few memories early on, and then more and more memories as they get older,' says Fivush. 'If you look at it statistically, below the age of three there's virtually no memories; below the age of seven or eight there are very few memories; and then there's a more consistent string of memories. Some of the more recent work has traced children longitudinally remembering: you ask a three- or four-year-old what they remember, and they will remember stuff that happened a year ago – but they won't remember those events as they get older. So a three- or four-year-old will remember things that happened when they were two, but by the time they are twelve they don't remember things that happened before they were three or four.'

Fivush doesn't think this is particularly strange. 'Forgetting happens: we lose those neural connections. Or, your memories aren't particularly important any more. It's not weird. Think about what would happen if you didn't forget anything! Forgetting is really adaptive.' Perhaps the importance of our early years is not in what we remember, but in the way these years set us up in terms of our brain development for our capacity to develop further, to regulate our emotions and interact with others later on. The pathways laid down in terms of brain development by early experiences have a far reach, even though we can't recall much of them.

## What happens to memory later?

After ten our working memory capacities stay fairly stable until about twenty-four or twenty-five, when we are at our cognitive peak in efficiency and speed of processing. If this is depressing, Fivush has some good news: 'We continue to use information pretty well, because we have all of this organized information to draw on.' The memory decline that we see in older adulthood is very variable, depending on the individual. 'Everybody loses the speed: they start having word-finding problems,' says Fivush. 'But do their memories become non-functional? I mean, a lot of older adults complain about memory problems – "I can't remember where I left my keys easily any more", "I can't remember this": that seems to be pretty common. But to the point of dementia, a lot of older adults have perfectly functional adequate memories well into their elder years.'

Rehearsal of memories helps too, and anything that we practise a lot, enough so it becomes automatic, can be held more in non-declarative procedural memory, like knowing how to ride a bike. This counteracts some of the effects of a decrease in efficiency that we might otherwise have as we age.

For most of us, though, when we think about memory, we tend to be thinking about our own memories of our lives. 'In talking in everyday language, when we say "I remember" we tend to mean autobiographical memory,' says Fivush: 'what I remember of my own personal experiences.'

## Autobiographical memory

Autobiographical memory is our own perspective on our lifetime's events, linked together in a personal timeline that develops as we grow and experience more and more. We tie our own memories of events together in this timeline with our sense of self: who we are is influenced by what has happened to us, and what happens next is influenced by who we perceive we are.

The role of autobiographical memory seems to go further than an automatic story book of our life so far. Having a personal history that we can recall gives us a sense of continuity and coherence, a way of defining ourselves that can be reassuring and is linked to our sense of self, or self-concept.[4] Our autobiographical memory informs our internal life scripts about what we should be doing, and connects us to our wider family network and their narratives. Thinking back to the chapter on attachment, our memories of our early relationships also form building blocks for the internal working models of relationships in the here and now.[5] How we react to others depends to some extent on how we remember and make sense of what happened with other people earlier in our lives.

A lot of the literature suggests that a key time for some of this creation of a personal narrative is adolescence. 'It's really in adolescence that we start to create an over-arching story of who we are,' says Fivush. 'Younger kids can remember their experiences, but they don't really – if you ask an eight-year-old, "Can you tell me your life story?" they tend to give you a few unconnected events. They remember – they have good memories – but it's not really connected into a meaningful life narrative.'

Many adults, asked to reminisce, tend to talk about their adolescence. This 'reminiscence bump' is a counterpoint to the period of infantile amnesia that we all have. Just as we all tend to forget these early infant years, so we are more likely to recall things from our teenage years. This is described more in Chapter 16, where the psychologist Catherine Loveday argues that this adds to the idea that teenage years are particularly important to us and our sense of self.

Autobiographical memory has been implicated in our ability to regulate ourselves. Being able to create a personal narrative of difficult experiences which is emotionally coherent, and makes sense to us in a way that isn't too self-blaming and which emphasises our ability for future control, is linked to better well-being, including better physical health, and better prognoses at school and beyond. How we process memories, make sense of previous experiences and are able to then set these aside to some degree, is extremely important for us and how we function going forward with our lives.

## Co-creating memories

My friend's three-year-old looks up from the juice she is drinking and says, out of the blue, 'Dancing'. We have not been dancing: we are in a café with her mother; we are not going dancing, or talking about dancing. 'That's right,' her mother says. 'We were dancing this morning. Who were we dancing with?'

Maternal reminiscing style, or the way that any important caregiver remembers shared past experiences with us as a pre-schooler, can vary profoundly, and has a big impact on our development of autobiographical memory. In studies that follow children over time, the style of a caregiver's reminiscing predicts a high degree of variance in child autobiographical memory, even after controlling for age, language, temperament and attachment.[6] Caregivers who used more positive, elaborative-reminiscing styles had children with more detailed and coherent personal narratives by the end of their pre-school experience. When the pre-school samples were followed up in adolescence, those who had caregivers with a more elaborative style of reminiscing reported earlier memories too.[7]

Elaborative reminiscing involves asking open-ended questions, and then integrating the child's responses into an ongoing narrative – just as my friend took her three-year-old's single word and then elaborated on it, helping her to create a larger narrative around who was dancing with her (her dad and sister in the kitchen). Elaborative reminiscing isn't necessarily to do with being more talkative in general: it's just a style of remembering, and it seems to be really helpful for children.

For Fivush this is one of the most exciting areas of development in the field of memory. 'I really think that my students and my colleagues who have worked on maternal reminiscing style have discovered something intriguing about how memory doesn't just develop in a social context but is also related to social and emotional growth.'

The research on caregiver-reminiscing style spans a broad range of areas: emotion regulation, conscience development, attachment ... One of those working on it has been Elaine Reese. 'She has got probably the longest longitudinal work data set on this early environment,' says Fivush, describing how Rees has looked at how early maternal reminiscing is influenced by and influences all kinds of aspects of both memory development and understanding of other people's minds and emotion.

Reese's work has robustly found that when parents talk about the past in a detailed, emotional and collaborative way, their children have stronger autobiographical memory skills. These are not the only benefits: children also are also better at language, and at describing and also regulating their own emotions, and paying attention.[8] There are some overlaps here with the idea of mentalisation from Chapter 7, and the idea that parents who can help their child to label their own internal world might also help them feel more emotionally contained, and able to manage their emotional ups and downs.

The power that a caregiver reminiscing with a child in an elaborative way about past events can have on many aspects of development is strikingly positive, both for children's potential and for the development of caregiver–child relationships. Several studies have now looked at the impact of brief training videos for caregivers on how to elaborate on their children's reminiscing, and have shown positive effects. Fivush is clear that not having had this isn't totally disadvantageous – other things can compensate – but it's such an easy point of intervention that it seems like a good one to run with, and anyone can do it.

Fivush draws parallels with literacy: 'Most parents want their kids to do well in reading, so they read to them, and we know that reading story books to your kids is a really good thing to do. But not everybody has easy access to story books, and not everybody is that literate and can really read that well to their kid. It turns out that talking about the past has some of the same effect as reading story books – maybe just as powerful. And that's an easy place: if you have a mother who doesn't have the educational skills or the resources to read story books, she can talk about her experiences. Everybody has that available, right?'

One group that might not is those children or adults who have grown up out of really difficult histories. Most work on reminiscing has been done on middle-class populations, but there is more research to be done with children from difficult backgrounds, and children of parents with traumatic histories. Fivush agrees: 'I don't think we know enough about the dark side of remembering.'

## Forming narratives of difficult experiences

This has overlaps with the wider literature on post-traumatic stress. For people who have experienced a specific traumatic event, whether as a child or an adult, there is a risk (though not a certainty) that they might go on to develop post-traumatic stress. This involves a triad of symptoms: avoidance, hyper-arousal and re-experiencing.

Avoidance is not wanting to think about the trauma, talk about it or go near the place where it happened. This is hardly surprising: it makes absolute sense that we wouldn't want to think or talk about something upsetting. Along with the other two symptom clusters, though, it can be problematic.

Hyper-arousal means being so hyper-anxious and stressed that a person is continually jumpy, feels the need to check behind doors or glance behind them a lot, and feels almost continually on high alert.

Re-experiencing involves flashbacks and intrusive thoughts: either having very strong memories, or actually feeling they are back at the scene of the trauma, having to experience it all over again.

One treatment for PTSD with a very good success rate is cognitive behavioural therapy, with an emphasis on 're-living' the traumatic event in a safe environment: going through the memories from beginning to end, with the aim of putting them in order more coherently, and challenging any self-blaming or shaming thoughts that come along with the memories. Taking the time to do this is the equivalent of emptying out and sorting a wardrobe full of clothes, so they no longer fall out on your head when you open the doors, but stay where they are until you go to get them on purpose. We want traumatic memories to be like this too: staying tidied away and not popping out when we least expect or want them.

Children who have grown up in a maltreating or traumatising environment are more likely to have a memory style which is over-general. Asking them questions about their early years results in less specific stories and more general replies, even when the questions themselves are quite specific. Over-general memory style is also linked with depression[9] and post-traumatic stress disorder,[10] so there is something about it that isn't helpful in the longer term. In the shorter term it might be protective:[11] why would we want to think about or remember things that are painful? It might also be that children who have experienced maltreatment and trauma haven't had the opportunities from the adults around them to help them talk about and scaffold the memories of their lives in a coherent way.

One of the things that talking therapies can do, thinks Fivush, is to help people re-work the narrative they have of their own experiences into something more helpful. 'It's not that you change what happened,' she says, 'but you try and change how that affects your current thinking. It's not just what happened, but how we make sense of it, how we understand that experience'. It might be that re-working a narrative can be done in other ways, too. A large body of research suggests that writing about stressful events can also have powerful effects on physical as well as psychological health. Again, there is something about putting together a narrative that seems to be helpful, whether spoken or written down.

With children who have experienced maltreating circumstances or caregivers, Fivush is clear that the impact on their autobiographical memories is complex. 'There's so many things going on with abused and neglected kids. I mean, cognitive development is generally delayed, both because they are not in a positive stimulating environment and because of other stress reactions related to chronic abuse which interfere with a lot of brain function. And it's probably the case as well that these mothers are not engaging in positive elaborative-memory talk.'

## Interviewing children

Everything we know about memory development in children tells us to tread carefully when we are asking children to remember things that have a lot riding on them. In everyday conversations it matters less if we accidentally sway a child's response to something, but in therapeutic conversations, or legal interviews, it's really important we make every effort to enable children to remember accurately.

Open-ended questions, that could be answered with any response, are better for getting accurate responses and not leading any specific answer, but younger children tend to give less information in response to open questions. Nevertheless, it's really important for police interviews with children, for social workers asking what might be going on at home, and for any professionals or personal acquaintances of children who are asking them about a potentially dangerous or criminal situation, that open questions are used. There is an effective ten-step protocol for helping children to relax and be able to recall specific events, developed by Michael Lamb in the US, and reviewed in a document on child-witness interview protocols by the British Psychological Society in 2007.[12] In the protocol he gets children to practise with a memory of something nice like a birthday party, before remembering the event he is asking about. He also goes through specifically what children should say if they don't know something, to minimise the chances of children going along with something that an adult is suggesting.

## Our memories and who we are

There is a long-standing consensus in both philosophy and psychology that personal identity is very strongly linked to our memories of our own experiences. This argument was made by the empiricist philosopher John Locke way back in the 1700s and still holds sway today: who we are does seem to relate to the memories we have of our lives so far. The meanings we make of new experiences are often largely influenced by this too.

Robyn Fivush agrees that our autobiographical memory is hugely important to our sense of who we are. 'It's not the only thing,' she says, acknowledging our physical sense of self and our more abstract self-concept – or idea of who we are. 'But when you ask people, "Tell me about yourself," they usually jump right in to telling you about their autobiographical history.'

On the other hand, Fivush is clear that we don't have to remember something for it to affect us. 'As adults we think we have to explicitly remember something, but we can't remember most of what happened explicitly: we can't recall it to mind – from the first three years of our lives. But the first three years of our lives set the stage, in terms of our understanding of relationships, and attachment relationships, and emotional bonds. We learn so much.'

We are also, I would like to think, more than the product of our experiences. Something about who we are dictates how we respond to experiences, so that even if two of us have had the same thing happen to us, we might make very different sense of it. Chapter 18 goes into identity development in more detail, but memory isn't the only thing that is added into the mix when we are thinking about who we are.

# 16

# Teenage Kicks

Teenagers get a bad press. Newspapers regularly connect teenagers with gangs in hoodies, anti-social behaviour and unplanned teen pregnancies. The stereotype is of 'Kevin' turning from sweet, polite child into grunting, grumpy thirteen-year-old on the dot of midnight. It's not uncommon in the UK for shops to limit the number of teenagers allowed in at one time. Teenagers who seem low or angry are often subject to having their moods explained away by hormones. Even with the best intentions, over-generalising is easy.

This isn't new. In Shakespeare's *A Winter's Tale* he writes,

> I would there were no age between ten and three-and-twenty, or that youth would sleep out the rest; for there is nothing in the between but getting wenches with child, wronging the ancientry, stealing, fighting.

Even Shakespearean teenagers were apparently lax with contraception and had violent and thieving tendencies.

What were you like as a teenager? Try to conjure up what that time felt like — what tastes and smells take you back there? Is it the aroma of Lynx and cigarette smoke? The taste of cheap alcohol and cherry lip balm? Or the smell of school corridors and taste of packed lunches? What about the teenagers you know now? Are they troubled, reckless and thoughtless – or are they actually quite good company?

When I think about my teenage years I can definitely identify some stereotypical moments: rule-breaking on school trips away; pretending to be old enough to buy bottles of cheap Lambrusco at the corner shop; a row with my parents about whether I could wear a certain pair of trousers to school . . . But loads of my teenage years were also really sensible: revising carefully for exams with endless revision cards; hanging out with friends in local cafes; worrying about how I came across to people . . . I sometimes felt more sensible than my mum and dad. The teenage stereotype is flimsy and two-dimensional. Like all stereotypes, it leaves a lot out.

## What is adolescence?

For a long time there have been more theories about adolescence than there have been experiments to test them. That is changing, and the last twenty years have seen a huge leap in what we understand about adolescent brain development and behaviour. Despite this leap forward, ideas vary about what adolescence is. No clear definition exists. We might argue that we don't even know what we're talking about.

For a start, trying to pin down the beginning and end of adolescence is elusive. We could define it using age, and say that adolescence is the teenage years, which is just what many health and education services do, but a twelve-year-old isn't radically different from a thirteen-year-old, and a nineteen-year-old doesn't suddenly get a download of maturity on their twentieth birthday. Having a cut-off like this might work pragmatically to keep boundaries to services clear, but it doesn't really tell us much about the nature of what's going on, and it doesn't account for very mature twelve-year-olds or younger-seeming twenty-year-olds.

We could go for physiological markers, like puberty or physical maturity, to mark the beginning and end of adolescence, but if we do that then the age bracket needs to go out the window, because the age at which these physical changes occur varies hugely between individuals, and can also depend on the environment individuals are in. Counter-intuitively perhaps, some studies have reported that people in harsher childhood environments tend to start puberty earlier – perhaps evolution's way of trying to encourage procreation sooner rather than later in an uncertain environment.[1] We'd also need to work out which physiological markers we thought were most important, because there are lots, and they don't all happen at the same time. Some of the hormonal changes we might expect to be related to adolescence actually occur very early, whereas some physical changes, like the pruning-back of un-needed connections in the brain, go on well into the mid-twenties.[2]

We could go for social markers instead, like the transition from primary to secondary school, or the ability to drive a car. Which ones to pick, though? We can legally learn to drive, smoke, drink, vote and have sex at different ages even in one society, let alone when we consider across cultures and continents. We might once have thought of the end of adolescence as being when we had settled down with a partner, ready to start a family of our own. Increases in student fees and difficulties in finding employment mean many young adults need to live at home for longer, and even later on it's no given that those life goals, especially starting a family, are present.

Age, physiology and social markers are all useful ways of thinking about the start and end of adolescence, and are all interlinked. Who develops first or grows the tallest is painfully obvious. Looking older than our years when we are in our teens can mean we get treated as though we are much older too, which, while it can bring some social advantages, can also leave us vulnerable. Being taller than our peers might leave us feeling self-conscious. Looking much younger can leave us feeling behind our friends, feeling like we'll never catch up. Physical changes affect us psychologically, and also affect the social opportunities available to us.

However we decide to define the exact beginning and end, most people, especially in Western cultures, think of adolescence as a stage we all go through as a prelude to and preparation for adulthood: a warm-up for the adult world. Having a warm-up is a bit of a luxury, and not always available to everyone, as we'll see later on.

## Stormy and stressful

One classic framework of what happens in this warm-up period is the theory of 'Storm and Stress': the idea that adolescence is inevitably tempestuous and full of turmoil because of baser instincts and drives that are raging around. This theory sees emotional crises as par for the course, all relating to drives for sex, violence, a jostling for independence and assertion of an individual identity, sometimes aggressively.

One of the main proponents of this theory, G. Stanley Hall (whom we met in Chapter 13 on play), was a psychologist from the early 1900s who became interested in child development, influenced both by Darwin's evolutionary theory and by psychoanalytic ideas popular at the time. He drew on ideas from Freud and from Peter Blos about how important it is to separate from our parents, and how the consequences of this can't be underestimated.

Blos[3, 4] called teenagers' move away from parents 'the second individuation process'. Along with several other psychoanalytic thinkers, he thought children first went through a period of individuation from their caregivers in the first three years of life. In adolescence, Blos thought, we need to separate more fully from our family and find a new identity for ourselves. He believed the loosening of the old familial ties resulted in an emotional void, an emptiness that needed to be filled – a hunger for new feelings, new experiences and new people. Seeking out new peer groups, trying out exciting things, having lots of group experiences and changing friendships often: all this, according to 'Storm and Stress' theory, fulfils that hunger,.

But along with the drive for shiny new experiences comes a sense of 'regression': going backwards to a younger, more uncertain state. At the same time that we as adolescents might want to push our family away, we also often feel we really need them.

This theory fits very well with the popular contemporary image of a typical teenager: a moody, risk-taking, sex-mad and aggressive bag of hormones in constant conflict with their parents and other authority figures. But does adolescence have to be this messy?

I certainly recognise that sense of ambivalence in wanting your parents to look after you and tell you what to do, but also really wanting them to stop giving you their opinion and let you get on with it. What strikes me is not so much that this isn't true for adolescence, but that it seems to me it could be true for most of us for quite a lot longer than our teenage years. When do we grow out of that odd tug-of-war of what we want from our parents? And why does it always get much worse around holidays like Christmas?

## Who the hell am I?

It wasn't just Blos, Freud and Stanley Hall who thought that adolescence was all about separating from our parents. Erik Erikson, another psychoanalyst by training, but one who forayed into other areas of psychology, thought that the main issue in adolescence was identity development. There's a whole chapter (18) devoted to how we develop our sense of self, but just as a footnote here, Erikson thought that all our life could be divided up quite neatly into various stages, one of them being the task in our teenage years of going from a state of identity confusion to one of identity formation. Finding out who we are, thought

Erikson, is usually preceded by an identity crisis, a period where we don't know who we are, what we are doing, or why. To his mind this could explain teenage moodiness and unpredictable behaviour.

Thinking back, how much does it seem true to you that you were trying to work out who you were as an adolescent? For me, I think I was certainly trying a few different ideas on, and I was definitely experimenting with a few different hair colours, but I'm not sure I got it all worked out by the time I left for university. In fact, I think I still sometimes try on different ways of being to see if they suit, and I've certainly been older than a teenager the last time I had a panic about which direction my life was going in. Chapter 18 goes into this in a lot more detail, but for now, let's just say it's a bit of an over-simplification to say that we get our identity all sorted out by the end of adolescence. Identity theorists now recognise that, while still seeing identity as an important theme of our teenage years.

These earlier ideas about what happens in the bit between childhood and adulthood came from people thinking about what made sense from a framework of unconscious drives and evolutionary explanations. The ideas weren't tested in the same way we might expect scientific hypotheses to be tested out today, and in fact the very nature of some of the ideas makes them hard to test. Motivational drives and feelings of emptiness aren't necessarily conscious, so it's hard to just make a questionnaire and hand it out to see if these ideas are true. It doesn't mean they're not valid, but it's a theoretical framework as opposed to a basis for experimenting.

## Changes in thinking

Imagine you had the option of being given a third eye. Where would you put it?

This is one of the questions thought up by Jean Piaget, the same researcher who was playing with marbles in Chapter 10. He asked this question of children of a range of different ages, and found that younger children, of about nine years old, would reliably say they would put it in the middle of their forehead. Older children, from about eleven years upwards, would have more inventive answers, putting the eye on their hand to see round corners, or in the back of their head to see what was behind them.[5]

Piaget saw this as one example of a qualitative shift in the way that children begin to think as they get older and enter adolescence. This was one example of a bunch of questions he asked children of various ages. He conceptualised the shift in thinking that he observed as a sign that adolescents reached a stage of cognitive development which he called the formal-operations stage. Adolescents become able to manage much more abstract thought, thought Piaget: manipulating 'contrary-to-fact' ideas, and considering things that are possible (or even absurd) rather than real. For Piaget, adolescence is the time that we become able to use hypothetical or conditional situations. This means that we can apply basic logic to problems which aren't concretely in front of us, problems which involve mental manipulations, conjecture and imagination.

Piaget thought the shift from having to draw or write out the different stages to a logic puzzle, to being able to think things through in our heads, means that familiar ideas that have already been learned can be re-visited and re-thought. This opens up a new independence of mind, and ability to question previously taught ideas. Piaget also thought that it is during adolescence that we are able to think much more about our own thinking, noticing thoughts and being able to evaluate them. For him, adolescence is also a period when we develop a different sort of time perspective, one which increasingly allows us to consider the future. These qualitative shifts in thinking abilities he saw as meaning that adolescents have a growing readiness to take on more adult roles in society.

Although some of these shifts in thinking are still attributed to adolescence, as with most stage theories in psychology it turns out that real life is a lot more complex than this neat picture of linear stages. Current research suggests that there aren't such discrete levels of development,[6] but that the emergence of different thinking abilities is more fluid. Not all young people even necessarily reach the formal-operations stage, and when we consider the complexities this stage involves, it's clear that some adults might also struggle with it. Thinking about thinking, for example, is something some people (certainly psychologists) spend a lot of time doing, but it doesn't come naturally to everyone. Young teenagers in particular actually need a lot more help with thinking about thoughts, with abstract reasoning and problem-solving, and with factoring the future into their decision-making. While Piaget undoubtedly contributed a lot to our understanding of cognitive development – he was right that the way we think does change as we get older, and our cognitive capabilities do increase – but advances in neuroscience now give a picture of considerable brain development into our mid-twenties. We know now that there is proliferation and pruning of brain cells[7] far beyond when we used to think brains were developing. We are also increasingly aware of the importance of brain development in regions related to social cognition, something which Piaget never knew about.

## Social media and selfies

When we think of adolescence today there is a whole fresh digital landscape to consider. Young people are flooded with different ways of interacting with each other, through words and pictures, 24/7. What used to be private has increasingly become public, from what someone's bedroom looks like to holiday swimsuit photos. For those who want to find them, highly self-conscious or comparative body shots are available, sometimes with invitations to 'rate' them. The hazard of private pictures being shared beyond their intended reach is commonplace. The limits of when this can happen have also been stretched: whereas the school day used to end and then you didn't have to see anyone until the next day, Facebook and other sites mean that interactions go on even overnight, and adolescents today can't be sure what they will wake up to.

The self-consciousness of seeing ourselves from the perspective of others has always been a focus in adolescence, but social media has brought an added intensity. It might be that a growing ability to think to some degree about our own thought processes during adolescence leads to an increased self-consciousness, and an increased ability to consider what other people are thinking about us.[8] What we don't yet know (although it doesn't stop lots of people having opinions on it) is whether the changing digital landscape of today's adolescence might also be affecting how adolescents think about themselves and others.

Research into the effects of the digital media explosion is still nascent – we just can't say with certainty what effect it has on teenagers' sense of self, although some studies suggest that selfies encourage low self-esteem, harsh evaluation and ideas of peers as judges.[9] It's easy to blame social media tools for the way they are used: for every sexy selfie that gets emailed round a school cohort there are many more completely innocent uses of Instagram or other sites. But growing research does seem to suggest that many of the uses of these new media can result in extra pressures, particularly relating to teenagers learning to see themselves as an object that is looked at and evaluated.

## Social learning

The effect of 'like's and 'dislike's and comments on social media posts is a prime example of social learning theory in action (see also Chapter 8 on people-training). Social learning theory is based on the idea that an individual's behaviour is continually shaped by their social environment and interactions.[10, 11] We tend to behave in a way that we see others behave in, especially as we're growing up, and we also respond to social rewards and punishments. If loads of people like a post that we write on Facebook we'll be more likely to write something similar another time, but if everyone ignores it we might think twice. In 'real life', if we continually get shunned by a peer group for behaving in one way, but get lauded for doing something else, it's more likely our behaviour will be shaped by the popularity of our actions. How we develop, throughout our whole lives, is affected by the interaction we have with our environment. This is even more true in adolescence, when we are especially sensitive to social rewards (but more of that later).

Whatever we think about digital interaction, most theorists agree that the most meaningful social interactions we have as adolescents are with our friends. The influence of the peer group we have is enormous. Where previously we might have looked to our families for approval, as we grow and begin to separate we seek a sense of self-esteem from those of a more similar age. The conflicting pressures of trying to live up to how we want our parents to see us, and how we want our friends to see us, can be huge. This is more complex now we live lives in parallel: with the people we are face to face with, and also with the communities we are digitally interacting with. Part of the task of adolescence is sifting through some of these conflicts and trying to hear our own voice in the chattering mix. What does my mum want? What do my friends think I should do? But what do I want to do – and what do I think is right? This is the stuff of values, of finding who we want to be. It comes with all the existential angst attached. Who am I? Can I make it on my own? What do I want to spend my life on? It's painful and difficult, and it doesn't go away at the end of adolescence, but it's also exciting and joyous: the stuff of independence; the stuff of freedom and choice.

## The importance of the social brain in adolescence

Thinking about ourselves in relation to others is something that takes up a lot of time as an adolescent. This doesn't make teenagers narcissistic or self-obsessed: it just means they are highly self-conscious and aware of how they measure up to their peer group. It's only with time and age that the excruciating self-comparison calms down, or is at least replaced by a more curious attitude towards differences between ourselves and others, rather than a desperate desire to be the same or to fit in. Even examples of non-conformist adolescent behaviour are often just a way of fitting in to an alternative tribe.

Older theories about adolescence have acknowledged the importance of the peer group, but in recent years brain-imaging research has pushed our understanding of this radically further. The new studies have shown that brain development goes on for much longer than anyone ever realised, and how vulnerable adolescents are to peer pressure and risk-taking in front of friends.

If I gave you the choice of £200 now, in your pocket, or £1,000 in six months' time, which would you prefer? Scientists at Temple University in the US asked this sort of question of nearly 1,000 individuals aged between ten and thirty, and found that the older someone was, the more likely they were to be able to delay gratification for a larger amount of money. Adolescents were more motivated by short-term reward than long-term reward, even when the future prize was much more substantial.[12]

The same research group[13] used a computer game, called the *Chicken Game*, to look at how adolescents' decision-making is affected by having friends nearby. They observed early adolescents (mean age of fourteen), late adolescents (mean age of nineteen) and adults (mean age of thirty-seven) playing the game. The game is a driving simulation, and the goal is to get your car as far along a course as possible, at the same time avoiding crashing into a wall which can appear at any moment without warning.

Participants were randomly assigned to play the game either alone or with two peers of the same age in the room. When they were tested on their own, participants in all of the age groups did a similar amount of risk-taking. When they were playing in a room with other similar-aged peers, the early adolescents scored twice as high on an index of risky driving, the late adolescents' driving was 50 per cent riskier, and the adults drove just the same whether peers were there or not. The younger teenagers were much more influenced by peer observation.

Studies like this suggest that teenagers tend to weigh factors differently from adults when they are making decisions. Where adults might consider future rewards and not pay much attention to other people's presence, teenagers don't take into account future consequences of their actions in the same way. They are more influenced by the social context in which they are making decisions.

Professor Sarah-Jayne Blakemore, a neuroscientist in London, got interested in these patterns of behavioural differences, and started relating them to changes in the brain development that functional magnetic resonance imaging (fMRI) was showing. Blakemore has spent much of her career looking at the changes that go on in our brains throughout adolescence, and is convinced that a specific time period is meaningful. While Blakemore doesn't want to propagate an unhelpful stereotype, she is curious about some of the neuroscience that might underlie and give rise to that stereotype: of the risk-taking, self-conscious and peer-led teenager. Her brain-imaging research shows that the areas of the brain that respond to rewards, and in particular social rewards, are hyper-responsive in teenagers – things that we all find reinforcing are *really* reinforcing for teens. At the same time, the areas of the brain most associated with planning, understanding other points of view and preventing inappropriate behaviour are the brain areas which are still developing in adolescents, and lagging behind adult-level functioning. Development of problem-solving, social skills and impulse control: all this is still ongoing. If teenagers are more impressed by rewards, especially social ones, and less able to fully think through a problem and plan solutions taking into account all points of view, it's not surprising that this might impact on the decisions they make, especially when they are hanging around with their friends.

So the decisions we made when we were teens are likely to have been swayed more by social pressures from our peer group, and by short-term gain. We would have been much less likely to factor in our twenty-year-old self (twenty seemed so far away!). When we consider the decisions many adolescents might be faced with, it's obvious how this might play out. A teenager trying to decide whether to lie to go out to a party tonight, even though it risks arguments and loss of trust later, is more likely to pick the immediately gratifying option that will gain social approval. Similarly with choosing whether or not to try smoking, or drugs, the immediate reward of the new experience and peer congratulation will be more influential than any longer-term downside.

'It's not that teenagers don't understand the risks,' says Blakemore: 'it's just that for some teenagers, in the moment, this understanding goes out of the window' – a sentiment probably all of us can relate to at some point.

## Reminiscing

Dr Catherine Loveday, a psychologist from the University of Westminster, has studied the tendency adults have to refer back to their teenage years when they are incited to reminisce. This 'reminiscence bump' adds to the idea that teenage years are particularly important to us. Loveday is clear in interview that she sees adolescence as a distinct time: 'I do think it's a very distinct period, biologically, socially, cognitively – I think all three of those things do go through measurable changes in that period.' As for the reminiscence bump, 'You can ask people for their favourite film, their favourite footballer, their favourite books – you can even just give them twenty words and say, "Tell me the first memory that comes to mind when I give you this word" – and still the bulk of those memories will come from that period. Even if you don't say, "Childhood sweets" – if you just say, "Name a sweet", or "Name a biscuit" – people will tend to come up with the stuff from that period. All these things illustrate just how clearly and easily we access that part of our memory bank. They're still the things that pop into our mind most easily.' For Loveday, this highlights the accessibility of this time of life in terms of memories, for the rest of our life. 'I would probably argue that that period of time shapes us for evermore, and we return to it for evermore for better and for worse . . . Even when you speak to someone of ninety they can usually access that period of time at the drop of a hat.'

## Blowing it all out of the water

We might have had flashpoints during our adolescence that we remember as being awful: heartbreak upon our first relationship ending; agony over how to handle situations in friendship groups; blazing rows with parents over whether we could go out or not – but how true is the idea that our entire adolescence is this roller-coaster ride? Surely much of it is also banal, boring and well behaved? Aren't lots of the teens we know really interesting to talk to, usually polite and sometimes shy?

Philip Graham, a child and adolescent psychiatrist and author of *The End of Adolescence*, argues that the idea of a stormy adolescence is a myth, and that the brain research about differences has been over-egged. He goes even further, to argue that there's not even any such thing as adolescence, that we have socially constructed the idea. He thinks experiments looking at differences between adults and teenagers have been over-generalised to fit a stereotype. He explained to me what he means by this. He doesn't think that teenagers make decisions in a different way, rather that they are making decisions about situations they've never been in before. He compares it to an adult learning to drive. We wouldn't expect someone who had never driven before to get in a car and drive off without any bumps. Similarly, for teenagers in new social landscapes, maybe they need some advice and gentle guidance. 'If they are moving into new types of social situation they do need more help with that.' Graham thinks the way that we think about teenagers underestimates them, and leads to negative expectations then creating problems. If you expect the worst, then that's what you get. Instead of chastising teens for 'risk-taking', Graham thinks we should think about it as learning and experimenting. He uses the example of a toddler learning to walk: would we call them a risk-taker? Or just understand that they aren't very good at walking because they haven't had much practice?[14]

There is, Graham acknowledges, 'a great deal of neuroscientific work suggesting that from about the ages of seven or eight, or maybe a bit later, there is an acceleration of a process of neuronal pruning, and that therefore the teens is a time of great intellectual development, and an opportunity that mustn't be missed to teach children things, because otherwise they will have missed out. It's a time when they are particularly receptive to ideas and to learning.'

Graham differs from Blakemore and others, though, in the importance he places on this: 'I don't see there as being a discontinuity from either earlier childhood or later into more mature adulthood,' he says. 'Indeed, as I frequently say, it seems to me that to accredit the teen years with a special status for learning – implying that if you haven't done it by the end of your teens you've really missed out – does a disservice to the fact that people continue to learn for the whole of their lives. People learn languages well into their sixties and seventies and eighties, sometimes with great facility. And all the other things which are said to be characteristic of adolescence, of those periods of the teens, seem to me not to be characteristic of that age period. Temper tantrums, for example, gradually reduce during the teens.'

Graham points to other civilisations where there isn't such a period of individuation: 'In societies where the process of detachment from your family never really takes place, you remain a close member of your family throughout your life – as far as women are concerned, until they marry and then they move into their husband's household. The idea in Western society that, yes, of course it is a period of increased independence: there's nothing physiologically determined about that.'

Adolescence, thinks Graham, is more of a rich, Western construct, dependent on resources. 'Even in non-human primates, whether the young go through a period of being allowed to lie around and laze and not do very much, depends on the availability of food supply. In the non-human or sub-human primates, if there's not much food around, there's no such thing as adolescence.'

## A longer adolescence

Just as some 'adolescents' in poorer societies, or societies where community values trump individual values, might not go through a period of practising independence in the same way, so too, in some contexts, we might carry on adolescence for longer. Teenage choices are supposed to be motivated much more by immediate social reward and less by longer-term negative consequences, but even for us as adults, many of our choices are still motivated by social reward – choices like whether to go on holiday with friends despite not really being able to afford it, or to go out and see friends three nights in a row rather than stay in and rest, or to drink too much at a party and be hungover the next day . . . Choices that take into account longer-term rewards instead of shorter-term gains are often hard ones to make. For every decision there is a loss, and when we are saying yes to an intangible future hope or dream we often have to say no to a shiny option right in front of our nose. It's hard making choices that aren't 'adolescent'. Even as a thirty-something-year-old. Maybe especially as a thirty-something-year-old.

The neuroscientific evidence, to me at least, is convincing enough to make me think we should make some allowances, and give teenagers some specific help with problem-solving and decision-making, as we know their brains are still developing in these areas, and that they might be unduly influenced by peer opinion. I don't think the social constructionist view should be ignored, though, and we should guard against getting too wedded to a modern-day type of stage theory. Just as Erikson's ideas about one identity crisis and resolution have turned out to be too simple, so adolescence is vague enough in its definition for it to be hard to draw out firm territorial boundaries. We all engage in risky or impulsive behaviour sometimes, even though our brains have presumably matured. The environment we are in, the cues we are being given by people around us, and the expectations we and other people have, make a big difference, and it's worth bearing that in mind.

# 17

# Pink and Blue

There is no society in the world in which it doesn't matter if you are male or female. In 2007 the World Economic Forum rated 128 countries on relative differences in economic participation and opportunity, how well people did in their studies, how empowered they were politically, and how healthy and long-living they were. The study found that, rated on these measures, men have a higher status than women in all the cultures they studied. What we have inside our pants is still an important predictor of status and well-being in societies worldwide.

Debates about whether these differences are socially constructed, or are a result of innate sex differences in behaviour and role, are continually present in Western media, often with no clear conclusion. What does the science of child development add to the debate?

## Gender and sex

Biological sex refers to the physiological characteristics that define men and women, and gender refers to the roles, behaviours, activities and attributes a society associates with men and women. Male and female are biological sex categories. Masculine and feminine are gender

Figure 17. Salary Differences Explained

categories. Gender is socially constructed, in the way we talk about men and women or girls and boys, and the different expectations we have of the different sexes. We can be feminine males and masculine females. Debates about biological sex and the gender someone feels they are have been increasingly opened up in recent years by the growing acceptance of gender identity disorder, and increasing treatment of this, at younger and younger ages. Individuals with gender identity disorder feel they have been born as the wrong sex: that their personality is at odds with their physical self.

How do we get a sense of what we are supposed to be like if we are male or female? Perhaps it helps to imagine the opposite: can you imagine having been brought up living a life where nobody knew what sex you were? Where whether you were a boy or a girl was totally secret? This is exactly how a Swedish couple tried to raise their child, whom they named Pop. In Toronto another couple did the same, trying to raise their child Storm as neither male nor female. Both sets of parents argued that they wanted their child to be free from gender stereotypes and expectations.[1]

Ideas about our gender identity are powerful, and they can affect what we think we can do. Other people's ideas about who we are can also affect what we think we can do. So ideas about gender roles aren't trivial.

Suggestions about the types of behaviour which are masculine or feminine can be passed on quite subtly. We might get them from the media around us, from the roles we see other men and women performing in our society, from the roles that males and females take on in our families. What are the stories in your family about which roles are taken up by men and women? What roles did your parents take on? Who was the breadwinner? Who cares for the sick? What are boys and girls supposed to do? Play nicely? Sit quietly? Be brave? Be strong? What about in the wider cultures that you come from? What is valued in men and women? Are the different sexes praised for the same things?

Looking at the characters that men and women play in films or have in books, and even how the different sexes are portrayed in advertising campaigns, can be revealing. One advert that was played in the cinemas for a well-known airline showed its magical employees realising from an early age that they had special powers. The females used these special powers to do things in their childhood like get the pair of shoes they wanted, and they then grew up to be the *crème de la crème* of air hostesses, with killer heels and lovely glossy lipstick. The men mostly grew up to be the pilots and engineers.

## Considering difference

These different depictions of men and women can either be seen as commenting on an inevitable truth, or as perpetuating a gender stereotype. There is something seductive about the idea of clear sex differences in behaviour. It's so much easier to use a broad-brush approach: 'Well, she's a woman – she would over-react', or, 'Well, he's a typical man: insensitive.' But how much does the evidence really stack up that boys and girls and men and women are so different?

There is a huge wealth of studies on children's behaviour related to what sex they are. You can find evidence for sex differences across all sorts of different domains, but these are often single studies which are not widely replicated.

Christia Spears Brown, Professor at the University of Kentucky, studies gender differences. Except the main finding is: there aren't many. Interviewed over email she explained: 'Based on large meta-analyses, which are large studies where they combine hundreds of other studies and look at all of the research together (for a total sample of more than 1 million people), we see that there are very few actual sex differences. Most of the differences we think exist between boys/men and girls/women are either non-existent or very, very small.

'For example, many people think girls are more talkative than boys. This is actually not true at all. Girls speak a few months before boys do, on average, but they end up talking the same amount. People also think boys are more aggressive than girls, but they are really only higher in unprovoked physical aggression. There is no difference in provoked aggression. That said, there are a few differences that do exist. In addition to earlier language development for girls and greater unprovoked aggression for boys, one actual difference is that boys, as infants, are a little more active and a little more impulsive than girls. Boys also throw a ball with greater velocity and power. There are then differences that appear as infants move into childhood. Boys have a better body image than girls, and boys have more confidence in their math ability, and less math anxiety, than girls (but they do not differ in math abilities or performance). That is really it for behavioural or personality differences.'

This is so different from the perceived wisdom of differences between the sexes. *Men are from Mars, Women are from Venus* would have had to have a whole new title if this was the accepted rhetoric: 'Men and Women, Both from the Same Planet After All'. It's maybe not as catchy, but it's much more accurate.

## Why do we see sex differences?

If there aren't such massive differences, then why do we think there are? Christia Spears Brown explains our tendency to see gender differences by innate bias: 'We all see what we expect to see. It is called the confirmation bias, and it just means we are biased toward seeing confirming information, and ignoring disconfirming information. This is why stereotypes are so hard to change. We only pay attention to behaviour that confirms the stereotype, and ignore the many, many times children do not fit the stereotype.'

Spears Brown thinks that this tendency to explain differences in characteristics and behaviour using sex differences can be even stronger when parents have a son and daughter: 'Any difference between the children is explained as a gender difference instead of an individual difference. It is easier to recognize differences as just individual differences when parents have same-sex children. For example, I have two daughters, so when they differ (and they always do), it is easier to recognize that the differences are because they are two unique individuals.'

This is pretty controversial, not just in everyday conversations but in the research community too. Despite the meta-analysis overview which shows the differences aren't consistent, there are still some research studies which do find differences. This might partly be to do with the tendency for academic journals to prefer to publish findings which show clear support of a predicted idea, rather than studies which show there isn't a difference. Just by chance it's likely that some studies will show gender differences, and they might be more likely to be published.

Even when studies are published, they aren't always clear-cut in what they say they show. A statistically significant difference between two groups doesn't mean it is a clinically meaningful significance. It just means it is unlikely to be caused by random variation in the sample. We need to think as well about how big the difference is,[2] and how much of the variance in whatever we are looking at is explained by that particular trait or characteristic – whether it is gender or something else. It's easy to get stereotypical quickly, when we actually need to carefully consider the evidence and its meaning.

## Are there *any* sex differences?

What about those differences that the meta-analysis did flag up? A tendency for girls to have lower body image and less confidence in maths . . . Spears Brown thinks these differences are largely socialized. 'Girls live in a world with a very thin body ideal (as is evident on every magazine cover in the supermarket check-out lines), and very few girls can live up to that ideal. They are also valued for their appearance much more than boys, and so self-esteem is tied to trying to attain an unrealistic ideal. In regards to math, boys seem to internalize the stereotype that boys are better at math than girls. Thus they are more confident and less anxious about their abilities. The irony is that their performance is actually no better than girls', and is often lower (especially in school grades). So boys are overestimating their abilities and girls are underestimating their abilities.'

By contrast, the unprovoked aggression difference is probably a combination of nature and nurture. 'Some people suggest that testosterone *in utero* plays a role in promoting aggression,' says Spears Brown. 'However, research with girls that are born with "boy levels" of testosterone does *not* show higher levels of aggression. Probably the testosterone lays the foundation, but then as a society we promote aggression in boys. Almost every toy and every video game that boys play with promotes aggression.

Not surprisingly, this heavy socializing is effective. The difference in throwing speed and force is also likely a combination of nature and nurture. Boys have a little more upper arm strength (and then go on to develop even more after puberty), but parents are more inclined to promote athletic prowess in sons than daughters. Lots of throwing sessions in the back yard are more likely to happen with boys, and improve those abilities.'

The only difference that seems to really be the result of 'nature' is the activity level and impulsivity of boys. 'This seems to appear in infancy, which suggests that it isn't the result of socialization. (Although we definitely foster and exacerbate these differences when boys are pushed more into high-activity sports.)'

## When does our awareness of sex and gender begin?

From the moment we are conceived as a pair of XX or XY chromosomes, our sex is determined, and we're cued up for a different menu of hormonal influences, *in utero* and beyond. It's the balance of androgens (male hormones) or oestrogens and progesterones (female hormones) which results in the differentiation of different physical characteristics and genitalia, and which some people use to explain sex differences in behaviour,[3,4] despite the meta-analytic evidence that these are over-egged.[5]

Our own recognition of sex happens surprisingly early. Studies[6] show that even by the age of one we already begin to have the ability to distinguish between male and female faces if there are gender-related cues, with long hair being associated with women and short with men. By two years old we begin to use 'he' and 'she' to label others, and by three we have a sense of our own gender identity, whether we are a 'he' or a 'she'. Between three and six years old we develop the idea of gender constancy – that is, the idea that whether someone is a 'he' or a 'she' is something that stays the same, doesn't change. If I am a girl, then I can't also be a boy. By three years old we not only recognise that there are different sexes, but increasingly prefer to spend time with children of the same sex as us. Stereotypes about gender also form early: ideas about clothes, play and behaviour are already around before six years old: 'What boys do' and 'What girls do' are ideas we get young, and ideas that most six-year-olds are already conforming to. One study showed boys tending to rate assertive character traits as the preferred way of being, and girls tending to value affiliative traits – i.e. where they are getting along better with others.[7] The effects of these roots, laid down so very early on, probably in the messages boys and girls have been given, can arguably be seen far in the future, in typical male and female styles of interaction at work and in relationships.

## Will you play with me?

Playmates at primary school, even if it's a mixed school, tend to be peers of the same sex, and boys and girls who play with the opposite sex risk being rejected by their own peers.[8] Observational studies suggest female children are more flexible in the type of play they get involved in, but in general show more self-control and pro-social behaviours than boys,[9] who tend to be more self-assertive, anti-social[10] and have a higher activity level[11] than the girls. These studies have small effect sizes, though, meaning the differences are not large, and they tend to study children in contexts where pressure to conform to a social stereotype is large, for example the playground. It's impossible to tease out the effect of the social messages that children will have already been exposed to about what boys and girls should do or be.

And children are definitely aware of gender differences. A US study showed that nearly all elementary school children know that all US presidents have been male. When asked why, most children referred both to some personal characteristics of men and also to gender discrimination.[12]

Even later on, in adolescence, when mixed-gender friendships get more common, same-sex friendships are still the most popular. Small but consistent differences in styles of relating are seen, with some studies showing that girls tend to talk about themselves and their own feelings more than boys,[13] and that boys, when they do talk about themselves and their feelings, tend to do this with girls, not boys.[14] One study found that adolescent girls tended to use supportive listening statements, a bit of a counselling technique, more than adolescent boys.[15] This could be seen as evidence of a gender difference in style, or it could be seen as evidence of a subtle but definite grooming of girls (not necessarily consciously) to take on more caring roles in society. Cultures that males and females are inducted into tend to involve more macho teasing for men and boys,[16] and more indirect forms of aggression, such as social exclusion or negative gossip, for girls and women.[17]

Which games children play, and which toys they use, have also been studied. A number of studies have tried to look at this, and some have found that the average play behaviours of boys and girls are different.[18] If small children are given five minutes to play, and are presented with a choice of toys (car, digger, ball, blue teddy, pink teddy, doll and cooking set), all placed equally far away from them, even nine-to-seventeen-month-old children show sex differences in play,[19] with girls spending significantly longer than boys playing with the doll, and boys spending significantly more time than girls with the car and ball.

Similar results have been found with young vervet monkeys,[20, 21] and much has been made of this. Some researchers suggest these findings point to an evolutionary explanation for these differences in how we play early on. All of these studies are only showing average differences, though. It doesn't mean that some girls don't go straight for the car and some boys don't love the doll. And although it's less likely to explain the monkey findings, young infant humans are very likely to have already been exposed to large amounts of social conditioning: adverts, family choice of toys, examples of peers playing with sex-stereotypical toys – so we can't necessarily infer that sex differences are biologically caused.

Spears Brown also remains unconvinced. 'I am never quite sure what to make of the primate studies,' she says, 'and really the toy-preference studies in general. I think they are a bit misleading.' She thinks the studies show small differences in the females' looking tendencies, and no clear trends in males. Even at a very young age it is likely that girls have been socialised towards female toys and, in the case of the primates, males also seem interested in dolls, so there is a real lack of clear evidence suggesting a strong gender difference. Even in studies where there is an effect, we don't know why.

## Everybody hurts

No matter what sex we are, we can experience life as difficult sometimes. The way that males and females tend to present to mental-health services as we are growing up does tend to show a slightly different trend. Boys report slightly higher self-esteem[22] during adolescence, and girls present with slightly higher incidence of depression[23] in late childhood and adolescence. Both sexes come up against media pressures to look a certain way, which have a negative effect on body image,[24] although in general teenage boys report better body image[25] than teenage girls. Body-image concerns can be exaggerated by same-sex peers,[26] and adolescence is a high-risk time for the development of mental-health issues related to body image, such as eating disorders or body dysmorphic disorder. Having said this, although body-image concerns might be an important trigger, eating disorders are rarely just about the food.

## All in the brain

The temptation to explain the small but consistent differences between the behaviour of males and females as based in the brain, rather than as a response to social conditioning, is huge, and whenever differences are found they tend to make headlines. One particular study in 2013[27] grabbed press interest when it claimed to find sex differences in the connectivity of male and female brains. The study used diffusion tensor imaging, a type of live imaging that allows white matter to be observed. It looked at nearly a thousand eighteen-to-twenty-two-year-olds, and compared male and female brains. The authors emphasised the tendency for males to have greater connectivity within each hemisphere of their brain, and the tendency for females to have more connections between their two brain hemispheres and across different brain areas. The conclusions the authors drew were that 'male brains are structured to facilitate connectivity between perception and co-ordinated action, whereas female brains are designed to facilitate communication between analytical and intuitive processing modes.'

This fits with the accepted narrative of male and female differences: that men might be somehow primed to be better at seeing threat and acting on it, while women analyse and intuit – but when the data is drilled into, this conclusion is hardly warranted. The overall sample the authors used was large, but they compared not only between the sexes but also between three different age ranges, compromising the size of the differences they could find because they looked at more variables. They gave little information on the participants, other than sex and age, so other confounding variables could have been responsible for the differences seen between the sexes, and they didn't look at the role of neurotransmitters in the brain, the potential for environmental effects on brain development, or the relationship between structural differences in the brains and any kind of functional difference on a specific test. Lastly, just because there were some differences in the samples, it doesn't mean that all men and all women are different in this way. Even if there is a statistically significant difference between the sexes there still can be a massive overlap between male and female results, and this means that many men and women will buck the trend.

We are far from being able to say that male and female brains are conclusively different, or from knowing what any differences necessarily mean. Our brains are not shaped like our genitals or coloured in pink or blue, however deliciously simple that might be if they were.

## Environmental influences: parents

The social influences on us as boys or girls come from all directions: family, friends, language and media. The script our family has about what men and women can do might not be conscious, but it might still be apparent, either in what mothers and fathers do for work or around the house, or through language that's used. Increased gender equality in family roles is associated with reduced gender stereotyping in children,[28] suggesting we learn something from what we see our parents modelling. In many studies, small average differences between motherly and fatherly parenting styles are seen in communication style, with mothers typically being more supportive and fathers more directive.[29] The ways boys and girls are played with also tend to be different,[30] with more rough-and-tumble play for boys when they are little, and later on more encouragement to be emotionally self-sufficient.[31] This trend towards fostering emotional autonomy in males is even stronger in studies of Asian and Latin American cultures,[32] suggesting it's more to do with cultural expectations of gender than with particular behaviour from children eliciting certain behaviours.

As children we don't just get ideas about how to be from our families: as we get older we spend more and more time with friends, and we look to our peers to model what we should be doing too. Gender-normative behaviour tends to be more likely in groups of one sex: for example, more assertive speech in all-male groups than in mixed groups.[33] This might result in a reinforcing cycle, with gender-normative skills being practised in same-sex groups and increasing the differences between the sexes, and perhaps also the likelihood to want to play separately.

In adolescence, body-image concerns can be exaggerated by same-sex peers,[34] and stereotypes reinforced in 'gender cultures'. Males tend to use macho teasing more than females, leading to inhibition of emotional intimacy,[35] whereas girls are more likely to use indirect aggression like social exclusion or negative gossip.[36]

## Language

The words we use day to day can contain powerful hidden (or not-so-hidden) messages about gender role and power. Using gendered nouns for job roles implies that they are single-sex positions: postman, chairman, headmaster ... There is a clear bias, then, that these jobs should be taken by a man. If that sounds daft, then imagine if we always used the female instead: postwoman, chairwoman, headmistress. It seems stranger, but there's no real reason why it should. Even neutrally pointing out that there are two sexes in a room, by saying 'Hello, boys and girls,' instead of 'Hello, children,' has an effect on behaviour, increasing gender stereotypes.[37] There are countless examples of words signalling a gender expectation – many partnered dance classes presume that men will be leading and women will be following; in the workplace many adult men (and some women) refer to grown-up women as 'girls'; even in high-status jobs women might be referred to with a term of endearment in a public meeting in a way that would just not happen to a man.

## Media

Media messages start early. Despite the rise of *Frozen* and other vehicles for strong female characters, TV commercials, children's TV, children's books, children's films all contain gender stereotypes that are so common as to be virtually invisible. Who is cast as rescuer or person needing rescuing? Who is given the lion's share of the action? Who is shown in pink or blue? The stereotypes range from big to small, but are never inconsequential. We are influenced by these messages as much as by ones in our real lives. If children see girls playing with only certain toys on TV, they will assume that these are girls' toys. The growing sexualisation of girls from a young age also has an effect. Images we see daily in music videos, cinema posters and TV ads show us airbrushed, semi-naked women whose bodies are fair game for comment. Clothes and stationery aimed at young girls have Playboy-Bunny symbols,[38] sparkly lips and sexy slogans. The messages out there about what is not only permissible, but even encouraged, are very different for male and female children. It is perhaps no surprise, although it is still shocking, that girls walking to school are cat-called by grown-up men in a way boys aren't subject to. Laura Bates, the writer and former actor, highlights similar examples on her website, Everyday Sexism. It is full of incidences not just from women but also from girls, all over the world.

## What can we do to help children feel freer from gender constraints?

Christia Spears Brown thinks there are three main things we can do: make gender less relevant to everyday life; correct children and other adults when they are falling into the trap of gender stereotypes; and focus on buying children toys that foster the traits we want to encourage, rather than buying toys that are supposed to be for a particular sex.

Changing the language we use, she adds, can also have a big effect on how gender comes into our everyday. 'We use it so often in our language ("What a smart girl!"), in the types of toys we buy, and in the colour-coding of everything in children's environment.' For her this continual labelling, colour-coding, and sorting along gender lines tells children that gender is the most important thing about them. 'Children then pay extra attention to gender, and foster their own stereotypes about what girls and boys are supposed to be like,' she says. 'Parents should think about whether labelling gender is really necessary in the situation or not. The answer is usually not. For example, instead of saying "the boys on the sidewalk", say "the kids on the sidewalk". This seems like a small and silly change, but the message is powerful, and backed up by research. Using gender in our language to label and sort people increases children's gender stereotypes.'

Being extra careful not to speak in gender stereotypes, and to point out gender stereotypes when we hear them, can also help children notice that these are stereotypes, rather than truths or universal expectations. Only setting an example isn't enough: Spear Brown advocates that we actively challenge.

By buying toys that don't fall into the gender divide we can also change stereotypes. This doesn't necessarily mean buying a bunch of guns for all our daughters. Spears Brown prefers to think about what behaviour we are trying to foster: 'For example, all people should be nurturing and empathetic. We foster it in girls, but not boys. This means boys should have dolls, stuffed animals and practise care-taking. All people should have good hand–eye coordination and spatial skills. This means that girls should play ball games and have building toys, such as Lego.' I certainly remember playing Meccano with my best friend David when I was little, and never thinking it was a boys' toy. He used to dress up in my nightie, too, and I don't think he felt that that was solely a girl's domain.

As adults, we can also think about how to liberate ourselves and each other from the limits of gender expectations. For Spears Brown 'the biggest issue is to recognise that gender tells us very little about what individual people are like. There are more differences between individual women and between individual men than there are between the average woman and average man. There is a full spectrum of gender expression that is perfectly normal.'

Figure 18. Me and David

## Gender fluidity and gender identity disorder

The idea of gender constancy – that our sex can't change – was for a long time seen as a developmental milestone, but in recent years this has been thrown into question somewhat. Gender fluidity is the idea that sex and gender are not fixed: that people might identify as male or female or neither – somewhere in between, or somewhere that changes day to day. While biological sex can't change easily, with advances in surgery and hormone treatment it can now change. How an individual chooses to dress or present themselves can also influence whether people around perceive them as male or female. The idea of gender fluidity is an emotive one, perhaps because our ideas about sex and gender are ingrained from a young age and bound up strongly with our ideas of identity.

Sometimes feelings about what sex we are can become so distressing that they really interfere with someone's ability to live their life, and this is when it might tip into gender identity disorder. Imagine feeling you were in the wrong body – trapped inside a sex you didn't ever want to be. Imagine being disgusted by your own genitalia, and constantly uncomfortable with the way you were expected to behave. Gender identity disorder is diagnosed when someone's gender identity is consistently and markedly at odds with their biological sex. It's nothing to do with sexuality – people with GID can identify as straight or gay: it's more to do with feeling at ease in a physical body. Treatment for GID is controversial because it can involve changing sex, or postponing puberty to allow the decision to change sex to be made later on. It's obviously a big decision to make, and no one wants an individual to feel they have made a mistake, so it tends to create a lot of anxiety in the systems surrounding young people who are reporting these feelings.

Not wanting to conform to society's expectations of your sex is different from GID, and the rise of gender fluidity can be seen as related to this. Christie Spears Brown thinks that gender fluidity is a complex issue, but not a new one. 'I think it is a reflection of how there are very few sex differences, and neither sex nor gender are the dichotomies we like to think they are,' she says. 'Determining someone's biological sex is complicated, and not always done accurately.' In cases where children have ambiguous genitalia or different chromosomal composition (e.g. not XX or XY but XO or XXY), then things might be more complicated. Different levels of oestrogen and androgens, the female and male hormones, can be seen in different individuals, and might have an effect on behaviour.

'Then add a very rigid culture in which only one type of girl and one type of boy is accepted,' says Spears Brown. 'Girls need to be feminine, into pink and princesses, quiet, passive, and polite. Boys need to be rough-and-tumble, highly active, aggressive, romantically interested in girls, and athletic. Any deviation from these norms leaves children feeling atypical. The reality is that most children don't fit neatly into these pink and blue boxes.'

Add to this sexual orientation – unrelated to biological sex, gender expression or gender identity, but part of the complex biosocial system we are all caught up in. Gender and sex are no longer clear dichotomies when we look at them like this, yet, as Spears Brown says, 'We presume boys and girls differ in many ways that they don't; and we have a rigid culture that punishes children who show any deviation from the presumed norm. It is not surprising that individuals find themselves moving across these different spectrums in various ways trying to make all of these components of sex and gender come into alignment.'

## Making sense of gender

Gender is a political issue, ultimately, as well as a personal one, and understanding how we develop our sense of self as a male or female could help us work better together in a society. Books like *Men are from Mars, Women are from Venus* start off trying to help, but in the end perpetuate the same tired stereotypes. There are some behavioural differences between the sexes, but these are often exaggerated, and most of the time there is more variation within groups of people of the same sex, on whatever trait is being measured, than between groups of people of two different sexes.

What differences there are can be explained using a mix of biological influence and environmental factors, and different arguments measure these out in different quantities. Biosocial theories that prioritise physical bodily differences suggest our roles are devised for our differing physical strengths: a capacity to fight, protect and hunt, and an ability to have children, nurture and rear them. This adequately explains history, for sure, but today, when many women choose not to have children, and in several contexts earn more than men, it's hard to subscribe to this as being an explanation we should necessarily stick to. Even biological drives, like an increased aggression associated with testosterone in some men, aren't seen in all, and arguably shouldn't necessarily be acted on.

Other theories place more emphasis on the thoughts we have about gender, which of course don't come from nowhere, but are shaped by the society we live in. Observations of role models, experience with play and learning, and direct teaching about what boys and girls should do: all this gets internalised, and these thoughts then shape our behaviour. This social cognitive[39, 40] theory nicely includes things like the layout of shops and gender cues present even in the aisles of the supermarket. It provides a theoretical framework for understanding that there are no gender-specific toys, only gender-specific marketing. Similarly, social-role theory also sees our thoughts as shaped by social cues, but focuses on roles in particular. The roles we see men and women take in our families, schools and workplaces shape our expectations of opportunities for development in work, in creative endeavours and in relationships. Roles give us expectations, but also opportunities to develop, and if those opportunities are limited from an early age we might not even be aware of those limits. If they are at odds with our natural desires we might begin to come up against them, in an unpleasant way. Just one example is the encouragement of assertive over affiliative behaviour that studies[41, 42, 43] suggest boys are exposed to over the years, which might be fostering a particular Western style of leadership or management quality and in turn placing such roles further out of the reach of women.

There isn't a consensus on the nature-versus-nurture debate on sex differences and gender. If anything, it's got more complex, with ideas about gender fluidity and gender identity disorder being more widely recognised. We are free to take our own position on how much we believe sex differences to be down to biology and how much to pervasive social cues, but unless we are die-hard biological theorists we need to pay attention to the evidence on the importance of the language and narratives we use around our children, and around each other. A swing in leadership literature towards valuing more traditionally feminine styles of compassionate leaders is opening up debate about different ways of working in higher-status roles, but the qualities needed to successfully reach those jobs are probably still traditionally masculine.

In a utopia, we would all be able to make peace with our nature whatever our sex, and decide we can be who we want to be, whether or not this fits with a social stereotype. Easier said than done, of course, but if we can't quite reach that, then at least being able to recognise our own biases and positioning in relation to the issues can be helpful. If we can hold our inherited beliefs up to the light, then there is some hope of us being able to challenge or champion them: to make gender politics explicit in the room; to allow for curiosity about other cultures, families, ways of being, and scripts about gender. Expectations about gender roles shape opportunities for development, and also create limitations. If one dominant voice is speaking, there is less room for others to be heard. The risk of entering the debate is to be labelled sexist in either direction, but the risk of not saying anything is a stultifying silence, and, even worse, a silencing of the least powerful.

## 18

# Who am I?

## The beginnings of identity

Draw a satisfyingly shiny blob of bright red lipstick on the nose of a child younger than eighteen months, and show them their reflection in a mirror, and they tend to reach out to touch the reflected red dot. Some of them even try to go behind the mirror, if it's free-standing, to search for the other child with the silly red nose.

Beyond about eighteen months, children behave differently. They reach to touch their own nose instead. They understand that the image in the mirror shows a reflection, and that they must be the little clown.

How do we go from reaching for the stranger in the mirror to understanding that it's us? And how do we go from there to being able to think about who we are and what we want to be? Identity is a slippery concept, and a very human thing to worry about. Who am I? What am I doing? What is the point of it all? These are excellent questions for rumination in a panic at 3 a.m., and they've also provided much material for philosophers, cognitive neuroscientists and psychologists, prompting a range of theories and experiments related to who we are and how we are able to think about this.

The 'rouge-on-the-nose test', with the lipstick-nosed babies, was first tried in 1972 by Beulah Amsterdam, who was copying similar experiments done in 1970 by Gordon Gallup with chimpanzees. Gallup anaesthetised chimpanzees and drew on their faces while they were asleep, an experiment that drunken students have been unwittingly emulating for decades. Researchers all over the world have used this test as an indicator of the first signs of a developing sense of identity, and it's an easy one to try at home if you happen to have some lipstick and a roving toddler.

Babies all develop at slightly different speeds, but in general the youngest that babies start to point towards their own nose when confronted with the rouge test is about fifteen months. By eighteen to twenty-four months most eighteen-to-twenty-month-old children start to use their own name, and the children who point towards their own nose also tend to use more personal pronouns, referring to themselves as 'I'. It seems that the rouge-on-the-nose test might be tapping into more than just an awareness of how a mirror works: rather a growing awareness of ourselves, a self-consciousness in the very basic sense of the word.

## What is identity?

The idea of identity is massive, and often nebulous: 'who one is, what one values, and one's future life course',[1] is how James Marcia, a renowned identity theorist, once described it. Not much, then. Some components of our identity are there from birth, though not necessarily unchangeable: our biological sex, our eye colour . . . Other aspects we decide later: our political preferences and favourite foods. Identity is multi-faceted, involving internal and external characteristics. We need to recognise that we are separate from other people, and be conscious that we can act in the world and influence it, which in turn affects how others might describe us. But identity also involves being able to think about our internal world, to have a sense of what we desire, prefer, believe and value.

## Why is it important?

Identity is rich fodder for all the psychotherapies in their different ways, and for good reason. Thinking about who we are might feel like navel-gazing, but not knowing who we are can be profoundly destabilising. We all have times when we might feel less clear about what it is we are doing in our life, but this can also be a major side-effect of mental-health problems.

Consider Mark, a teenage boy who loves to play basketball, who plays in a band with his friends, who dreams of being famous. He also hears voices. They started a year or so ago and have got steadily worse, getting in the way of his concentration at school, making him feel irritated and on edge, saying things that make him feel sad. He had to defer his first GCSE exams, and he's not sure if he'll be able to take the rest. His sense of who he is has been blown apart by his diagnosis of psychosis. He doesn't want to think about it like that; it makes him feel worse. It makes him feel like one of those people they write about in the papers: someone dangerous. He doesn't feel dangerous: he feels really scared.

For Mark, he hasn't yet had enough experiences of success to feel he has a reliable narrative of who he is. While experiencing mental-health problems at any point is potentially destabilising and distressing, for younger people it has even more potential to knock their sense of who they are. If Mark had started hearing voices ten years later, after having already passed all his exams and obtained a record deal, he would have a different sense of self to fall back on, and with which to make sense of his experiences.

While even when we are adults our lives can always travel forward in many different ways, if we have fewer personal successes and experiences we can remind ourselves of, then anything which shakes our sense of who we are can potentially have a bigger impact. As a child and teenager we have certainly had less chance to explore the possibilities of who we are. Big life events, in particular physical or mental illness, can have a big effect on how we see ourselves, especially when we are too young to have an established personal story.

## Developing identity

How do we get to this sense of who we are? How do we go from being babies – bundles of desires and frustrations – to grown-ups with at least some sense that we can choose, change, direct where we are going?

Developmental psychologists have tended to cluster aspects of identity into self as subject, and self as object. Self as subject means understanding ourselves as someone who can know things and act in the world: the main character of our own plot. It involves having a sense of personal agency or capability, a sense of distinctness from others, a sense of continuity of self, and a sense of self-awareness.

Self as object is how someone else might describe us, if they were watching our story unfolding. This includes a sense of how we would describe our own attributes and roles, a sense of who we are in terms of physical and psychological characteristics. How would your best friend describe you, if they were sitting here now?

Our knowledge of who we are as subject or actor develops before our ability to describe ourselves as an object, or as seen by someone else. Before we can know what we are like or how someone else might see us, we first have to know that we exist at all. As young infants, we develop our sensory capabilities, our motor skills and our ability to influence the world around us, including other people. If we think about the development of what we can see and hear, and what we can do with our bodies (see Chapters 3, 5 and 6), we can imagine how developing an ability to grab things that we want, or observe the world more clearly, might impact on how we understand who *we* are. A sense of ourselves as a separate person able to act in and on the world develops first, followed by a growing understanding of what we are like and what we are good at.

If you had to describe yourself in five words which words would you pick? Take a moment to think about it. Chances are your responses will be pretty sophisticated. It's likely you will have thought about something to do with your character or personality. When children are asked this question their response tends to vary predictably, depending on what age they are. Pre-school children tend to talk about characteristics you can see. They might mention hair colour, pets, or behaviours they have learned to do, like counting.[2] Their tendency is to talk about concrete physical things or clear behaviours, not psychological tendencies: 'I have a cat', for example, or, 'I have green eyes', or, 'I can say my alphabet.'

This isn't to say that pre-school children can't reflect to some degree on what they are like psychologically. Many experiments with children yield richer results when an effort is made to make the verbal load in the experiment smaller – that is, to say things more simply, and give clearer options. Given some help to understand and think about it, pre-school children are capable of stating preferences which reflect their personalities – but it's not what they naturally go for when asked about themselves. R. A. Eder and colleagues[3] gave children a forced-choice task, asking them to choose between statements like 'It's more fun to do things with other people than by myself' and 'It's more fun to do things by myself than with other people.' Pre-school children gave consistent responses to repeated questions like this, suggesting they weren't just picking options at random.

In middle childhood we start to get a bit more naturally reflective. We tend to use more trait-based words like 'helpful' or 'naughty'. At this age we tend to relate our character to other people as well, either by describing how well liked we are – 'I have lots of friends' – or by comparing ourselves to other people in our family or in our class: 'I'm more hard-working than Tom', and 'I'm louder than Jay.'

As teenagers we get even more sophisticated. Values, political and religious views, abstract ideas about what we believe in: these things start to be part of the discussion. As adolescents we are able to think about how we might be different in different social contexts as well: 'I'm loud when I'm with my family, but when I'm with my friends I'm one of the shyer ones.' This sort of difference in who we are and how we behave can be really confusing, especially for young adolescents.

Adolescence has long been labelled as a period of identity development (see Chapter 16 for more on teens). Erik Erikson, an identity theorist and developmental psychologist with a psychoanalytic background and a name like a Swedish pop star (who we met in Chapter 16), came up with an overarching theory of how we all develop throughout our lives. His model suggests that we move through a series of life stages, each one involving a crisis between two states of being that needs to be resolved before we can move on to the next. Erikson was the first person to use the term 'identity crisis', a term we now all use fairly flippantly. Erikson was referring to the crisis between identity and identity confusion that he thought occurred for all adolescents. According to Erikson, as adolescents we are supposed to be figuring out what is unique about ourselves in order to be able to choose our roles in life.

## Identity in relationships

'Success' in working out who we are is also supposed to be associated with intimate romantic relationships later on. Not having a clear idea of who we are is thought to make it harder to sustain a close relationship with someone else. In a line reminiscent of a Spice Girls lyric, Erikson wrote that 'the condition of twoness is that one must first become oneself'.

Is this true? Do we need a strong sense of who we are in order to be able to have an intimate relationship? Wim Beyers and Inge Seiffge-Krenke from Germany carried out a study looking at this in 2010. They followed nearly a hundred adolescents, fifty-two girls and forty-one boys, over ten years, giving them questionnaires and interviewing them to measure their sense of self and their attitudes to intimate relationships. They started at age fifteen and followed up at age twenty-five.

The questionnaire which measured sense of self in the fifteen-year-olds was the Washington University Sentence Completion Test, which has a series of sentence beginnings which participants complete: for example, 'When I am criticised . . .' and, 'My mother . . .' These get coded according to how much they relate to a very self-oriented attitude, or whether they begin to take into account other people's viewpoints more, which the authors think suggests a stronger sense of self. The Network Relationship Inventory, the questionnaire that measured attitudes towards intimacy at fifteen, asked questions such as how much secrets were shared with a trusted other. Ten years later, the participants were interviewed and rated on how coherent a sense of identity they had, and how much they were able to be intimate with, committed to, but also respectful of their partner's autonomy.

The researchers found that scores on sense of self at age fifteen strongly predicted intimacy in relationships ten years later. Individuals with weaker scores on sense of self either didn't have enduring romantic relationships, had more superficial romantic relationships which didn't involve much emotional intimacy, or were in what the authors referred to as 'merged' relationships, where they compensated for anxiety about the relationship by being overly close but couldn't really say what they felt.

This is only one study, but it echoes other findings, and it uses a nice follow-up design. It makes sense that the stronger our sense of self, the more able we might be to have a truly intimate relationship, where we are able to say what we really mean, and also tolerate hearing what the other person has to say without our sense of who we are crumbling around us. It seems to support Erikson's idea that developing a sense of identity lets us get closer to other people.

## Identity or intimacy: which comes first?

To play devil's advocate, does identity really come first? It's likely to be much easier to develop a sense of who we are if we have first experienced a good enough intimate relationship with our initial caregivers. Everything we thought about in Chapter 2 suggests that we are set up to develop a more reflective, coherent sense of who we are when we have had early caregiving that is stable, reliable and emotionally responsive. In order to develop a sense of self we are helped by having a more secure early experience of being held in mind. Intimacy and identity might help each other to develop.

## Is identity that stable?

Erikson's big theory of a series of crises does have a reassuring air of living a life like a level-based computer game – levelling up after each crisis is resolved, even if a series of inevitable crises sounds a bit stressful. But it definitely has its flaws, too, and Erikson himself acknowledged and re-worked it towards the end of his career. A major problem is whether the development of a sense of our identity really is this simple. Do we all end up with a coherent sense of self that we develop as a teenager and that then remains constant? If my identity had been formed by the time I was seventeen then I would still be a vegetarian, I would be both less careful and less confident, and I would be more open to wearing black lipstick.

James Marcia built on Erikson's theory of identity development, but questioned the idea of a stable identity. He thought our sense of identity continually changes through our whole life, or gets 're-formulated'. Although it might be important to have a clear idea of who we are as we transition to adulthood, this is still subject to getting knocked about a bit, and it can change, sometimes radically.

Marcia's way of explaining identity development uses two dimensions: exploration and commitment. He thought that our identity status fell somewhere along each of these dimensions: the extent to which we have explored, questioned, thought about who we are and what we stand for, and the extent to which we have committed to one particular version of ourselves.

He uses four categories: identity diffusion, moratorium, identity foreclosure and identity achievement. Obviously one of these options sounds much better than the others, and I think it's fair to say Marcia felt that identity achievement was where we wanted to end up. He had a more flexible view of how we get there than Erikson's original ideas, though, and he allowed for us to cycle in and out of our 'achieved' identity.

The four different categories fall at different points along the two dimensions. Identity diffusion refers to low exploration and low commitment – a person, for example, who has not explored very many possibilities related to who they might be, but who has also not really committed strongly to any sense of what their values are. For instance, they might be non-committal about what their views are on politics, but from a standpoint of not really bothering to read the papers or talk to their friends about it at all.

Moratorium represents a large amount of exploration and low commitment: a period in our lives where we might be trying a lot of things out but not feeling fixed on any of them. Trying out lots of different types of music, or lots of different sorts of fashions, might be a good example of this: one weekend a goth, the next into hip hop.

Identity foreclosure is having a high level of commitment to a certain sense of identity, but without having explored many other options – maybe just going along with our family's view of who we should be. Being totally sure of a political standpoint, for example, without having found anything out about other perspectives; or having a strong narrative of there being a certain way to do things as a family, without ever having tried anything different: 'We don't like to travel too far afield on holiday – that's just how we like it.'

Finally, identity achievement is coming from a position where several options have been explored, but there is also a strong sense that certain values and beliefs have been stuck to. 'I tried working privately, but it just didn't suit me: I like working in teams more, and I really believe in public healthcare, even if it's more stressful and worse paid.'

Questions that researchers used from the earlier studies on identity achievement tended to cover occupation, religion and political beliefs, for example:

How willing do you think you'd be to give up your job/religion/ political ideology if something better came along?

Example classifications of the answers were:

*Identity diffusion*: Oh, sure. If something better came along, I'd change just like that.

*Moratorium*: I guess if I knew for sure I could answer that better. It would have to be something in the general area.

*Identity foreclosure*: Not very willing. It's what I've always wanted to do. The folks are happy with it, and so am I.

*Identity achievement*: Well, I might, but I doubt it. I can't see what 'something better' would be.

Later writings, by Jane Kroger in 2007,[4] described identity achievement as 'elusive', with only half of adolescents in most samples reaching what Marcia would classify as 'identity achievement'. Whether or not we agree that we reach a sense of identity in our teens, Marcia thinks this sense is still elusive in adult life. Marcia thought that as adults we flip between moratorium and achievement, with major life events tipping us off-balance, but then a sense of identity being reconstructed. He thinks many different events have the potential to tip us off-centre, even if we have a strong sense of who we are.

## Can happy events knock our sense of self?

According to Marcia, the events that make us question who we are don't have to be crises like a mental-health difficulty or a major loss. He thought 'disequilibrating events' were anything unsettling, and that any of these can knock our sense of equilibrium about who we are. These events can be positive things, too, like falling in love or changing career path: things that might make us happy, but can also unsettle our sense of who we are.

Falling in love is a really interesting one. Have you ever had that sensation of feeling out of touch with yourself when you start to find yourself very attracted to someone else? Schools of thought on falling in love cluster into those who think it is so destabilising that it's more like having an addiction, and those who think it's a valuable experience that helps us to develop a richer sense of ourselves. In reality it's probably both, and Marcia's theory allows for this. A happy event, but one that promotes change, so it's both disturbing and fulfilling.

Much of this literature assumes that having a relatively stable identity and a long-term relationship are the optimum ways to live a life. Of course, there are lots of people who don't want to have just one intimate relationship, and who want to experiment more with who they are in life. Does this mean they don't achieve identity development? Again, Marcia leaves room for this, and his idea of cycling between moratorium and achievement (or MAMA as he calls it) is a more fluid way of thinking about who we are. It's a bit more hopeful when things go wrong as well. For Mark, the teenager we thought about earlier, perhaps his psychosis is a period of crisis which prompts moratorium, but this doesn't prevent him reaching a sense of identity achievement.

## Identity and brain

Identity is such a complex concept you'd think it would be hard to find a clear brain area that relates to its development.

This hasn't stopped scientists from trying, and finding some promising leads. The main brain structures implicated in the development of our sense of self are called the cortical midline structures. These are in the middle of the brain, clustered around the corpus callosum, the inverted-banana-shaped structure that runs from front to back between the two hemispheres of our brain.[5]

The most common way of looking for the brain regions involved has been functional magnetic resonance imaging (fMRI), a brain-scanning technique which can measure brain activity with relatively good spatial and temporal resolution while we are lying in a brain scanner doing tasks.

Lisette Van der Meer and colleagues reviewed the adult imaging studies done up to 2010 and analysed them together They compared the brain scans of people who were lying in the scanner thinking about something related to themselves, with baseline scans of people not thinking about anything in particular.[6] Then they compared brain scans of adults who were asked to think about characteristics of themselves (self-reflection), with brain scans of people thinking about characteristics of other people (other reflection). They pinpointed one main brain area that differed when we think about ourselves instead of someone else. This was the ventral medial prefrontal cortex, part of those cortical midline structures. They concluded that it is this bit of the brain that is involved in reflecting on ourselves.

But when children and teenagers are put in the brain scanner and compared on similar tasks, the results are slightly different from the adult findings. For example, when nine-to-eleven-year-olds are asked to rate themselves on how good they are with friends and at school, and then asked to rate Harry Potter on the same things, several parts of the brain are more active during the self-reflection task, and they aren't the same region as in the adult brains above.[7] By contrast, adults asked to do the Harry Potter experiment did show more activation in the ventral medial prefrontal cortex in relation to thinking about themselves, just like the adults in the studies reviewed by Van der Meer.

This suggests that children and teenagers might be thinking about themselves in a different way to adults, literally using a different area of their brain. One thing the authors of the Harry Potter experiment suggest is that children and teenagers might have more of a social idea of their identity, so thinking about things to do with their own skills might require the brain areas involved in processing information to do with other people. Maybe as teenagers what we think of who we are is more intertwined with how we think other people see us. This would fit with the idea of a developing self-identity that is more socially malleable when we are teenagers. One of the nice things about getting older is that feeling of, 'Well, this is who I am: take it or leave it.' Perhaps the different brain-scan patterns show us this in action.

Ultimately this is still at the stage of conjecture, though, and we still know very little about how the landscape of the brain promotes our understanding of who we are. It's unlikely we are going to find one part of our brain which is responsible for our sense of self: it's much more likely to be a network of areas which is involved, including those in the adult and child brain scans described above, but probably also some others too, not only within the cortical midline structures.[8]

## Who is that in the mirror?

Next time you look in the mirror or catch a glimpse of your reflection on the Tube, give a moment's pause to consider how sophisticated this moment of cognition is. You know the reflection is you, you probably have some value judgements about how you look related to how you want to present yourself to others, you have some idea of how you look now in relation to how you used to look when you were younger.

And you have a sense of being able to act in the world to change how you look, what you do, even to some extent who you are. To some degree we are all constantly developing our own identity through what we choose to focus on thinking about, what we choose to do, and how we choose to look. That we can think about all this gives us a uniquely human perspective, which sometimes leads to painful dilemmas and those middle-of-the-night crises, but also gives us the chance to really choose whatever for us is 'a life well lived'.

# 19

# Fully grown?

From before birth right up into adulthood, we've done a whistle-stop tour through some of the key areas of child development, and how these different aspects affect our adult selves.

## Fancy a re-cap?

We started way back in Chapter 1 considering that period of time before we are even out there in the world. From the darkness of the womb where we first develop from a bundle of cells into a fully formed foetus ready to be born, we grow in our abilities to respond – to sounds, to our mother's emotions, to our own personal in-uterine environment, and, if we are twins or triplets, to the only other being or beings who will ever experience this particular environment with us. Birth is a massive shock: out of the womb now, we are in the world, with whole realms of experiences to adapt to and learn from, our sight rapidly catching up with our hearing, and the foundations being laid in those first two years for all sorts of things: our speech, our abilities to regulate our own emotions, our patterns of relating to our caregivers, and later on to other people who are close to us.

The caregiving we receive in those first early years sets us up for interacting with others throughout our lives; Chapter 2 dug right down into this. The overlaps between attachment styles with our parents and patterns in relationships with other people we are close to later on, including romantic partners, mean that the psychological concept of attachment relationships can be a continual source of fascinating potential insight into ourselves and how we are. Every time I read about or lecture on attachment I feel I am learning something new. Chapter 9 was the painful counterpoint to the chapter on attachment, considering what happens when care isn't quite good enough, although the hopeful strand in this was that even if our primary caregivers have struggled to provide care that was good enough, an important relationship with another caring individual can do a lot to help.

In Chapter 3 we thought about the milestones that new parents, and to some degree children, get bombarded with, and how useful or not this can be. There are some milestones it can be helpful to know about, and some amazing ways of experimenting with babies and children to understand the development that goes into these milestones being achieved. Remember the experiments which measure where babies look, from Chapter 5 on visual development? Or the *Neighbours* theme tune experiment from Chapter 6 which was about the development of speech? Having said that, many milestones can also be taken with a pinch of salt, especially those relating to what we 'should' all be doing in our adult lives. Another perspective we challenged was the nature-vs-nurture debate, with the epigenetic advances that Chapter 4 describes giving us a whole new link between upbringing, environment and genetic predisposition, and adding layers of subtlety to our understanding of biological effects.

The ability to take up other perspectives was also something we chewed over in Chapter 7. The development of theory of mind (seeing the world from another's point of view) and the ability to mentalise, or to see ourselves from the outside and others from the inside, are crucial to our ability to interact with one another. Mentalising goes wrong many times a day for all of us, but it's a good state to be striving for: trying to see other people's perspective on things, or at least being curious enough to properly ask. The opposite is also important: taking time to fully explain what we are thinking and feeling, and not assuming that someone else will instinctively know.

Ideas from behaviourism, about how to encourage some behaviours more than others, were explored in Chapter 8, in relation not only to toddler management but also to how we encourage ourselves to stay on track with doing what we plan, and how to interact with others in a way that means we don't keep accidentally encouraging behaviour from others that makes us feel bad.

In Chapter 10 it all got a bit meta-cognitive, when we were thinking about thinking. Remember Piaget and his stage theories? And how we saw that his theory has been upgraded with a more fluid approach, incorporating multiple types of intelligence, and including the role of other people in helping us to reach our maximal potential, stretching us through the zone of proximal development?

We thought, too, about moral development, in Chapter 11, and how we also learn a lot of this, although the role of emotion and the newer evidence from babies and toddlers suggests there is also a biological basis for some of our moral instincts, and one which, chillingly, not everyone possesses.

Development doesn't happen solo, and we thought a bit in Chapter 12 about the effects our family has on us, in particular our siblings or our lack of them. The effects aren't clear-cut: there is no easy sibling-related fortune-telling here, but in any case that would be way too blunt, and would rob you of the chance to quiz yourself on what you think the effects of your sibling context were on you.

Whether or not we had siblings, we hopefully all had friends, and the way we play, as children and as grown-ups, with friends and on our own, is as important to our child development as it is to our continued adult well-being. Chapter 13 was all about playtime, before in Chapter 14 we went back to school, considering the different educational contexts and our interaction with them, as children, possibly as parents, or as professionals working within or with the school system.

Throughout the whole book I've been asking you to remember different snippets of your own childhood, and sharing some memories of my own. The very capacity for this sort of reminiscing is what we thought about in Chapter 15, with Chapter 16 then catapulting us smack-bang into memories of our teenage years, along with the different theories people have about the point of adolescence, the changes which go on, and whether indeed the concept even exists.

From here we went to gender development and gender roles in Chapter 17, and while I could probably fill a book with ideas about gender alone, we left this after a mere chapter to get properly existential in Chapter 18 about our identity as a whole – who we are, how this links to our past experiences, and what it might mean for our futures.

We arrive now, at the end of this tour, back here with ourself, the same self who started this book, but maybe, hopefully, with some new knowledge and some fresh ways of looking at who we are and where we have come from.

## Making sense of the past

The Roman God Janus has two heads, one looking to the past and one to the future, to symbolise new beginnings, transitions, passages and endings. The Sankofa bird from Ghana is drawn looking backwards, to symbolise learning from the past. It's quite a decision: choosing to rummage around in past experiences with an idea that this will shed some light on what's happening now, or give us a clue about what to do next. I hope this book will provide some overview of the big ideas in child development, and let you have a think about how some of these might be relevant in some ways for your own life.

## *Plus ça change, plus c'est la même chose*

Despite our capacity for continual change and evolution, it's usual for us all to have moments in our lives where we have felt a sense of being in a repeating pattern of some kind, and not always in a good way. A romantic relationship that started off so well and yet somehow ends in the same old way as so many before; an argument that feels the same even though it is with a different person each time; a sense that we are sabotaging ourselves at work in the same ways again and again; a familiar reaction to an unfamiliar set of circumstances . . . We might even know what we are doing that we want to change, but it happens again anyway.

If it were as simple as thinking back and having a light-bulb moment about 'daddy issues' or 'being an older child' then we would clock it. But it's not simple. Human development is incredible. We grow from a ragtag bundle of cells into the most sophisticated beings on the planet. With that sophistication comes complexity. We aren't always logical. We don't always behave well. As adults we might have more capacity to control how we react to the emotions we feel inside, but not always, and even more so when they remind us, consciously or unconsciously, of scenes from our childhood, relationships, emotions, connections.

Sometimes we are so ingrained in our habits, or those around us are, that the change can be a new understanding rather than a radical shift of behaviour – that slight shift of perspective, as if a kaleidoscope has been twisted just slightly, and a whole new set of colours and shapes has been released into place. For some this unlocks a new sense of self-compassion, or compassion for others. If we can understand behaviours that we or those around us engage in, then we can sometimes feel softer about them. It doesn't have to stop us trying to change our reactions or lay down clear boundaries about what isn't OK for us, but it might help us do this in a more loving way.

We don't necessarily need to have therapy for these sorts of insights to happen, but some of the ideas from different therapies can also be helpful to know about.

It's not for everyone, or right for every time, but many talking therapies draw on psychological theories which acknowledge, to some degree or other, that we are products of our own previous experiences, and that in order to move forward we sometimes need to understand what has gone before, and have the opportunity to have something different.

## Therapies: what's on the menu?

The different schools of talking therapies are based on different core theories and draw on different evidence bases, but there are also enormous overlaps.

### Hot cross buns

Cognitive behavioural therapy (CBT) links together thoughts, feelings, behaviours and bodily sensations, in a model often referred to as a 'hot cross bun'.

This way of working is based on the idea that if we change just one of these things, then the rest will change too. If we alter our thoughts, then we can change how we feel; if we do something differently, then our thoughts might shift; if we feel different emotions, then our bodies will feel different physically.

### Third wave

So-called 'third-wave' CBT includes therapies which have grown out of cognitive behavioural theory, but taken its ideas in slightly different directions.

### Dialectical behavioural therapy (DBT)

This is a tricky mouthful to say. As a therapy it focuses on trying to help people to accept the impossible dilemmas we face in life, and

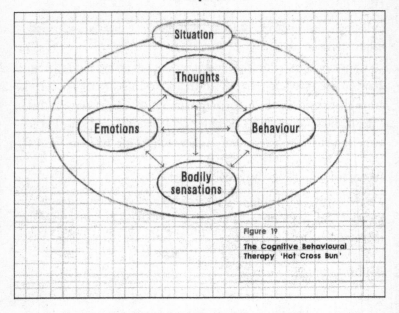

Figure 19. The Cognitive Behavioural Therapy 'Hot Cross Bun'

accept who we are and how we tend to respond to these dilemmas. That isn't to say that we then change nothing – but that we both accept that it feels impossible to change at the same time as learning skills to help us to be able to. In particular, these skills focus on understanding and managing moods, and finding ways to soothe ourselves rather than hurt ourselves in times of difficulty. DBT is particularly useful for individuals who might use self-harm to cope with massive ups and downs in emotion.

### Acceptance and commitment therapy (ACT)

ACT pays particular attention to the core values lying underneath our behaviours and decisions. It gets us to think about what it is that we really want from life, and what we want our actions to be guided by. Particularly helpful when faced with decisions or dilemmas where we feel we don't know how to find a way out of them, ACT tries to get us to tune in to our basic guiding principles, and then use these to steer the direction we are headed in. Instead of thinking, 'What should I do from other people's perspective?' or, 'What is the right thing to do?' it liberates us to think about 'What decision would be best for my guiding principle of wanting to prioritise family?' or, 'What decision would be the best one in terms of my guiding value of creativity?' It might seem like a small shift, but it can lead to big differences.

## Compassion-focused therapies

These do what they say on the tin. They prioritise compassion. Compassion is the recognition that someone is in pain or suffering, and the desire to want to alleviate that suffering. We can be compassionate to others around us, and also to ourselves. Compassion-focused therapies try to help us to take a compassionate stance towards both other people and ourselves, and it's often the latter that is hardest. We are often extremely good at criticising ourselves, beating ourselves up, and hitting ourselves over the head with a bunch of 'should haves'. Trying to let go of this, and regard ourselves with the same understanding and care that we might a really good friend, can be incredibly difficult, but can reap significant benefits.

## Mindfulness-based therapies

Being in the here and now, rather than tangled up in worries about the past or about the future, is much easier said than done, but is one root of many meditation practices from around the globe. Mindfulness-based therapies have recognised the benefits of meditation and tried to draw and build on them, incorporating them into broad cognitive behavioural frameworks to help with anxiety, depression, stress reduction and other specific difficulties.

## Under the surface

Psychodynamic therapies place more importance on the unconscious. A bit like how an iceberg floats about with most of its mass being underwater, these therapies see the conscious thoughts we have as being the mere tip of what we are feeling, wanting and processing. The early experiences we have are seen to be pivotal in the creation of our unconscious frameworks for life and relationships. Therapy is a chance perhaps to make some of the unconscious conscious, but not even necessarily this: it might be to have a different experience of relating with the therapist, which can lead to alternative unconscious maps of relationships that can be used going forward – quite literally expanding the unconscious terrain. It's not a short process, and the very nature of it is that we may not be aware of the benefits for some time either.

## Human systems

Systemic therapies consider the networks of inter-relating people that we are part of. They prioritise thinking about our effects on one another, and the functions of behaviours as part of a system; about the roles we are invited to take on within our networks that might relate to wider networks and cultures; and about the stories we are handed by other members of the systems we are in, and how they influence us. Like a row of dominoes, nothing we do is without an effect on those around us, and we cannot help but be influenced by the beliefs and actions of those same people. Family and systemic therapists often invite wide family networks to attend therapy sessions, but can still work with individuals as well.

## Story-weaving

One therapy that has grown out of systemic theory is narrative therapy. This pays particular attention to the stories we tell about ourselves and our lives. The stories we create about ourselves often have a powerful effect on what we feel able to do next. Like drawing lines between stars to create a constellation, to some extent we create our own picture by choosing which stars, or events, to join together. I could tell a narrative of being a great driver, passing my test first time, never having a crash, and this might make me feel more confident in taking to the wheel. Or I could tell a narrative of not having driven for several years, and of my mum feeling the need to put imaginary brakes on whenever she is a passenger with me. This might make me less inclined to rent a car.

When we prioritise events which tell a problem-saturated story, it can be hard to escape this. Narrative therapists help us to spot exceptions to the story and generate more of these, running with and strengthening an alternative narrative which might then free us up to be able to do more things going forward.

## Same-same or different?

There are other therapies, too: cognitive analytic, which lands in the middle of CBT and psychodynamic, and schema-focused therapy, another third-wave CBT, as well as many others, but these are some of the main routes that therapy takes.

Different talking therapies suit different people, and some have a better proven track record with certain specific difficulties. One way of spotting the difference between the theories has been to use a framework which the systemic therapist John Burnham called 'approach, method, technique'. This will be familiar to students of education, as it's a framework sometimes used to think about language learning. Burnham uses it to think about therapy style, and the theory behind why different therapists do therapy differently. 'Simply put', explains Burnham, 'approach concerns why something is done, method how it is done, and technique with what is done'.[1] The different types of talking therapy can differ widely in why they are doing what they do, and in what this looks like.

Some of the differences in theoretical approach are bigger than others. A therapy which prioritises the unconscious is likely to look quite different to one which prioritises making conscious links between thoughts and feelings. They might both be aiming to help solve similar problems, but their way of going about this could be radically different. Their approach to the effect that our childhood might have on us as grown-ups is also likely to be different. Both might recognise that our childhood landscape gives us building blocks for later adult thoughts and experiences, but the theories behind this might be hugely divergent.

Whatever the focus of the therapy, I don't think there is a single talking therapy which doesn't acknowledge that our early lives have an impact on us as adults. Whatever legacy we were given, whether it was darker or lighter, or somewhere in the middle, we all have dramas of childhood and experiences of solace which were present when our brains were still forming. This is not to say that the effects of this are totally set in stone. It is never too late to make a change, gain a new understanding or have a different experience, even a small one. We are continually evolving in response to the people and environment around us.

## Are we there yet?

So just when are we a grown-up? How do we know? I feel a range of different ages even throughout one day, and the extent to which I feel equipped to deal with the adult world also varies. The current social and economic landscape has some bearing on when we can consider ourselves adults, too, in terms of which adult milestones are within our reach.

Perhaps we are never fully grown. There is always room for more growth, change, expansion into new parts of ourselves and shedding of others. Life is continually changing around us, and as the demands made of us shift, so we have fresh opportunities to respond in new and different ways. Looking back and getting stuck in 'what ifs' is not helpful, but if we can use psychological theories as a scaffold to support our thinking, then we might be able to use the views we have of the past to give us clues for how to read the map ahead. With a theoretical framework to help, we can look back and understand some of what went on before. This can, on a good day, free us up to understand the bits that stay the same, and perhaps even allow us to sidestep or tweak the course of the trajectory we have been set on, if we want to.

# Dedication and acknowledgements

I've loved writing this book and thinking about how childhood makes us who we are. How my childhood made me who I am was largely down to my mum and dad, Barbara and Piers. This book is dedicated to you both, with lots of love.

I think we carry on growing and changing throughout all of our lives, and how my adulthood so far has made me who I am is also down to a fantastic family of friends and loved ones who give me a web of connection and support that I am hugely grateful for. Thank you.

The work we do and the teams we work in also shape us, both professionally and personally, and I have been lucky enough to have learned a massive amount from clients and colleagues past and present. Special thanks to Dr Sophie Browning and the rest of the team at Snowfields Adolescent Unit, and to colleagues and guest lecturers at the Anna Freud Centre where I got the idea to write this book. I interviewed many of the guest lecturers and have also drawn from ideas presented in other talks that I sat in on. Thanks also to Dr Jane Mellanby, Dr Chris Barker and Lee Smith for formative influences and encouragements at different points.

# Dedication and acknowledgements

This is the first book I've ever written, and it turns out that the moment where you hand the book in is really just the start of a big old process involving many other people. Thank you to the team at Little, Brown and to James Wills from Watson Little for all that you've done.

# Notes

## Chapter 1

[1] P. G. Hepper, 'Foetal "soap" addiction', *Lancet* (11 June 1988), 1347–8.

[2] P. G. Hepper and B. S. Shahidullah, 'The development of fetal hearing', *Fetal and Maternal Medicine Review*, 6: 03 (1994), 167–79.

[3] To put this in context, as adult humans we can hear between frequencies of 20Hz and 20000Hz, so our hearing is better than when we are a foetus, although as we get older the top end of the frequencies we can hear decreases again.

[4] K. M. Godfrey and D. J. Barker, 'Fetal nutrition and adult disease', *American Journal of Clinical Nutrition*, 71: 5 (2000), 1344s–52s.

[5] J. H. Kim and A. R. Scialli, 'Thalidomide: the tragedy of birth defects and the effective treatment of disease', *Toxicological Sciences*, 122: 1 (2011), 1–6.

[6] D. A. Frank, M. Augustyn, W. G. Knight, T. Pell and B. Zuckerman, 'Growth, development, and behaviour in early childhood following prenatal cocaine exposure: a systematic review', *Jama*, 285: 12 (2001), 1613–25.

[7] M. A. Huestis and R. E. Choo, 'Drug abuse's smallest victims: *in utero* drug exposure', *Forensic Science International*, 128: 1 (2002), 20–30.

[8] Samuel Taylor Coleridge (1772–1834).

[9] National Child Welfare Educational Poster for mothers, 1919.

[10] M. L. Schneider, C. F. Moore, G. W. Kraemer, A. D. Roberts and O. T. DeJesus, 'The impact of prenatal stress, fetal alcohol exposure, or both on development: perspectives from a primate model', *Psychoneuroendocrinology*, 27: 1 (2002), 285–98.

[11] B. R. Van den Bergh, E. J. Mulder, M. Mennes and V. Glover, 'Antenatal maternal anxiety and stress and the neurobehavioural development of the fetus and child: links and possible mechanisms. A review', *Neuroscience & Biobehavioral Reviews*, 29: 2 (2005), 237–58.

[12] Interview with Vivette Glover.

[13] N. Bayley, *Bayley Scales of Infant Development* (2nd edition, San Antonio, TX: The Psychological Corporation, 1993).

[14] J. A. DiPietro, M. F. Novak, K. A. Costigan, L. D. Atella and S. P. Reusing, 'Maternal psychological distress during pregnancy in relation to child development at age two', *Child Development*, 77: 3 (2006), 573–87.

[15] J. L. Cox, J. M. Holden and R. Sagovsky, 'Detection of postnatal depression: development of the 10-item Edinburgh Postnatal Depression Scale', *British Journal of Psychiatry*, 150: 6 (1987), 782–6.

## Chapter 2

[1] For a video look up: http://www.youtube.com/watch?v=eqZmW7uIPW4&feature=related

[2] http://www.youtube.com/watch?v=KlfOecrr6kI

[3] M. D. S. Ainsworth, 'The development of infant–mother attachment' in B. Cardwell and H. Ricciuti (eds), *Review of Child Development Research*, vol. 3 (Chicago: University of Chicago Press, 1973), 1–94.

[4] J. Bowlby, *Attachment. Attachment and Loss*, vol. 1, *Loss* (New York: Basic Books, 1969).

[5] https://thepsychologist.bps.org.uk/volume-22/edition-10/looking-back-making-and-breaking-attachment-theory

[6] L. A. Sroufe and E. Waters, 'Attachment as an organizational construct', *Child Development* (1977), 1184–99.

[7] M. Main and J. Solomon, 'Discovery of an insecure-disorganised/disoriented attachment pattern' In T. B. Brazelton & M. W. Yogman (eds), *Affective development in infancy* (Westport, CT: Ablex Publishing, 1986) 95–124.

[8] M. H. Van Ijzendoorn and P. M. Kroonenberg, 'Cross-cultural patterns of attachment: a meta-analysis of the strange situation', *Child Development* (1988), 147–56.

[9] L. A. Sroufe, 'Attachment and development: a prospective, longitudinal study from birth to adulthood', *Attachment & Human Development*, 7: 4 (2005), 349–67.

[10] M. J. Dykas, S. S. Woodhouse, K. B. Ehrlich and J. Cassidy, 'Do adolescents and parents reconstruct memories about their conflict as a function of adolescent attachment?' *Child Development*, 81 (2010), 1445–59.

[11] E. Waters, S. Merrick, D., Treboux, J. Crowell and L. Albersheim, 'Attachment security in infancy and early adulthood: a twenty-year longitudinal study', *Child Development*, 71: 3 (2000), 684–9.

[12] P. M. Crittenden, 'A dynamic-maturational model of attachment', *Australian and New Zealand Journal of Family Therapy*, 27: 2 (2006), 105–15.

[13] C. M. Pistole, 'Attachment in adult romantic relationships', *Journal of Social and Personal Relationships*, 6 (1989), 505–10.

[14] J. A. Lee, *Colours of Love: An Exploration of the Ways of Loving* (Toronto: New Press, 1973).

[15] P. R. Shaver and C. Hazan, 'A biased overview of the study of love', *Journal of Social and Personal Relationships*, 5: 4 (1988), 473–501.

[16] D. Marazziti and S. Baroni, 'Romantic love: the mystery of its biological roots', *Clinical Neuropsychiatry*, 9: 1 (2012), 14–20.

[17] R. Feldman, I. Gordon, I. Schneiderman, O. Weisman and O. Zagoory-Sharon, 'Natural variations in maternal and paternal care are associated with systematic changes in oxytocin following parent–infant contact', *Psychoneuroendocrinology*, 35: 8 (2010), 1133–41.

[18] T. R. Insel, 'A neurobiological basis of social attachment', *American Journal of Psychiatry*, 154: 6 (1997), 726–35.

[19] C. A. Pedersen and A. J. Prange, Jr, 'Induction of maternal behavior in virgin rats after intracerebroventricular administration of oxytocin,' Proceedings of the National Academy of Science USA, December 1979, 76: 12, 6661–5.

[20] https://www.ncbi.nlm.nih.gov/pmc/articles/PMC3312973/

[21] http://www.nature.com/news/gene-switches-make-prairie-voles-fall-in-love-1.13112

[22] https://www.ncbi.nlm.nih.gov/pmc/articles/PMC3312973/

## Chapter 3

[1] Oxford English Dictionary online (2007).

[2] Philippe Ariès, *Centuries of Childhood: A Social History of Family Life* (New York: Alfred Knopf, 1962).

[3] H. Montgomery, 'An introduction to childhood', *Anthropological Perspectives on Children's Lives* (UK: Wiley-Blackwell, 2009).

[4] A. Gottlieb, 'Non-Western approaches to spiritual development among infants and young children: a case study from West Africa', *The Handbook of Spiritual Development in Childhood and Adolescence* (2006), 150–62. Discussed by Taiwo Afuape in lecture at Anna Freud Centre in 2014.

[5] J. J. Campos, B. I. Bertenthal and R. Kermoian, 'Early experience and emotional development: the emergence of wariness of heights', *Psychological Science*, 3: 1 (1992), 61–4.

[6] T. N. Carraher, D. W. Carraher and A. D. Schliemann, 'Mathematics in the streets and in schools', *British Journal of Developmental Psychology*, 3: 1 (1985), 21–9.

[7] B. Carter and M. McGoldrick, *The Changing Family Life Cycle: A Framework for Family Therapy* (Boston, MA: Allyn & Bacon, 1989).

[8] S. Timimi, 'The McDonaldization of childhood: children's mental health in neo-liberal market cultures', *Transcultural Psychiatry*, 47: 5 (2010), 686–706.

[9] https://www.nice.org.uk/guidance/cg72/chapter/recommendations

## Chapter 4

[1] There are some exceptions, but this is the majority of cases.

[2] Hypothalamic Pituitary Adrenal axis – a major neuroendocrine system which controls stress reactions and regulates many body processes.

[3] D. Marr and T. Poggio, 'From understanding computation to understanding neural circuitry', Artificial Intelligence Laboratory, A.I. Memo (1976), Massachusetts Institute of Technology.

[4] C. B. Pert, *Molecules of Emotion: Why You Feel the Way You Feel* (Simon & Schuster, 1997).

# Notes

## Chapter 5

[1] J. Van Hof-Van Duin and G. Mohn, 'The development of visual acuity in normal fullterm and preterm infants', *Vision Research*, 26: 6 (1986), 909–16.

[2] T. K. McLellan, 'What Can Your Baby See? Parenting', http://www.parenting.com/article/what-can-your-baby-see – retrieved online 3 October 2017.

[3] R. J. Adams, M. L. Courage and M. E. Mercer, 'Systematic measurement of human neonatal color vision', *Vision Research*, 34: 13 (1994), 1691–701.

[4] D. Y. Teller, D. R. Peeples and M. Sekel, 'Discrimination of chromatic from white light by two-month-old human infants', *Vision Research*, 18: 1 (1978), 41–8.

[5] L. Hainline, 'The development of basic visual abilities' in A. Slater (ed.), *Perceptual Development: Visual, Auditory and Speech Perception in Infancy* (Hove: Psychology Press, 1998), 5–50.

[6] R. L. Fantz, 'The origin of form perception', *Scientific American, 204,* issue 5 (1961) 66–73.

[7] R. L. Fantz and J. F. Fagan III, 'Visual attention to size and number of pattern details by term and preterm infants during the first six months', *Child Development* (1975), 3–18.

[8] R. L. Fantz and S. B. Miranda, 'Newborn infant attention to form of contour', *Child Development* (1975), 224–8.

[9] R. L. Fantz, 'The origin of form perception', op. cit.

[10] D. Maurer and M. Barrera, 'Infants' perception of natural and distorted arrangements of a schematic face', *Child Development* (1981), 196–202.

[11] C. C. Goren, M. Sarty and P. Y. Wu, 'Visual following and pattern discrimination of face-like stimuli by newborn infants', *Pediatrics*, 56: 4 (1975), 544–9.

[12] M. H. Johnson, S. Dziurawiec, H. Ellis and J. Morton, 'Newborns' preferential tracking of face-like stimuli and its subsequent decline', *Cognition*, 40: 1 (1991), 1–19.

[13] Simion et al., (2001); think the preference for face-like stimuli is due to an innate preference for face-like qualities. F. Simion, V. Macchi Cassia, C. Turati and E. Valenza, 'The origins of face perception: specific vs non-specific mechanisms', *Infant Child Development*, 10 (2001), 59–65.

[14] Johnson (1998) proposed two separate processing systems:

> Early processing system to attend to moving faces;
>
> Second processing system to identify static faces
>
> (cf. dorsal/ventral streams hypotheses of vision: object recognition and object for action).

[15] G. E. Walton, N. J. A. Bower and T. G. R. Bower, 'Recognition of familiar faces by newborns', *Infant Behavior and Development*, 15: 2 (1992), 265–9.

[16] O. Pascalis, S. de Schonen, J. Morton, C. Deruelle and M. Fabre-Grenet, 'Mother's face recognition by neonates: a replication and an extension', *Infant Behavior and Development*, 18: 1 (1995), 79–85.

[17] Janine Oostenbroek et al., 'Why the confusion around neonatal imitation? A review', *Journal of Reproductive and Infant Psychology*, 31: 4 (2013), 328–41.

[18] A. N. Meltzoff and M. K. Moore, 'Imitation of facial and manual gestures by human neonates', *Science*, 198: 4,312 (1977), 75–8.

[19] M. Anisfeld, G. Turkewitz, S. A. Rose, F. R. Rosenberg, F. J. Sheiber, D. A. Couturier-Fagan, J. S. Ger & I. Sommer, 'No compelling evidence that newborns imitate oral gestures', Infancy, 2 (1), (2001) 111–22.

[20] U. Castiello, C. Becchio, S. Zoia, C. Nelini, L. Sartori, L. Blason ... and V. Gallese, 'Wired to be social: the ontogeny of human interaction', PloS one, 5: 10 (2010), e13199.

[21] R. D. Walk and E. Gibson, 'The "visual cliff"', Scientific American, 202: 4 (1960).

[22] A. N. Schwartz, J. J. Campos and E. J. Baisel, 'The visual cliff: cardiac and behavioral responses on the deep and shallow sides at five and nine months of age', Journal of Experimental Child Psychology, 15: 1 (1973), 86–99.

[23] Campos, J. J., B. I. Bertenthal and R. Kermoian; 'Early experience and emotional development: the emergence of wariness of heights, Psychological Science, 3 (1992); (1), 61–64.

[24] R. D. Walk, E. J. Gibson and T. J. Tighe, 'Behavior of light- and dark-reared rats on a visual cliff', Science 126, issue 3263 (1957) 80–81.

[25] C. Blakemore and G. F. Cooper, 'Development of the brain depends on the visual environment', Nature, 228: 5,270 (1970), 477–8.

[26] M. Otten, A. K. Seth and Y. Pinto, 'A social Bayesian brain: how social knowledge can shape visual perception', Brain and Cognition, 112 (2017), 69–77.

[27] E. Anderson, E. H. Siegel and L. F. Barrett, 'What you feel influences what you see: the role of affective feelings in resolving binocular rivalry', Journal of Experimental Social Psychology, 47: 4 (2011), 856–60.

[28] E. Anderson, E. H. Siegel, E. Bliss-Moreau and L. F. Barrett, 'The visual impact of gossip', Science, 332: 6,036 (2011), 1446.

## Chapter 6

[1] J. Mehler, P. Jusczyk, G. Lambertz, N. Halsted, J. Bertoncini and C. Amiel-Tison, ' A precursor of language acquisition in young infants', Cognition, 29: 2 (1988), 143–78.

[2] The distinct units of sound in a language which distinguish one word from another, like 'b', 'p', 'd' and 't'.

[3] V. Fromkin, R. Rodman and N. Hyams, An Introduction to Language, 10th edition (Wadsworth: Cengage Learning, 2014).

[4] Fromkin, Rodman and Hyams, An introduction to Language, op. cit.

[5] W. O'Grady, How Children Learn Language (CUP, 2005).

[6] Fromkin, Rodman and Hyams, An Introduction to Language, op. cit.

[7] Fromkin, Rodman and Hyams, An Introduction to Language, op. cit.

[8] Noam Chomsky, The Human Language Series, Part 2: 'Acquiring the Human Language' (1994). Available at: http://thehumanlanguage.com/films/two/

9 Deb Roy, 'The birth of a word', TED talk (2011):http://www.ted.com/talks/ deb_roy_the_birth_of_a_word

## Chapter 7

1 Some people argue that the children are just confused by the question, which admittedly is an odd one – why show someone a pencil and then ask what someone else would think was in there?

2 B. M. Repacholi and A. Gopnik, 'Early reasoning about desires: evidence from 14- and 18-month-olds', Developmental Psychology, 33: 1 (1997), 12, and discussed by Dr Iroise Dumontheil in lectures at Anna Freud Centre.

3 R. Baillargeon, R. M. Scott and Z. He, 'False-belief understanding in infants', Trends in Cognitive Sciences, 14: 3 (2010), 110–18, http://doi.org/10.1016/j.tics.2009.12.006

4 I. A. Apperly, 'What is "theory of mind"? Concepts, cognitive processes and individual differences', Quarterly Journal of Experimental Psychology, 65: 5 (2012), 825–39.

5 T. T. Rivet and J. L. Matson, 'Review of gender differences in core symptomatology in autism spectrum disorders', Research in Autism Spectrum Disorders, 5: 3 (2011), 957–76.

## Chapter 8

1 E. T. Gershoff, 'Corporal punishment by parents and associated child behaviors and experiences: a meta-analytic and theoretical review', Psychological Bulletin, 128: 4 (2002), 539.

2 L. J., Berlin, J. M. Ispa, M. A. Fine, P. S. Malone, J. Brooks-Gunn, C. Brady-Smith, C. Brady-Smith and Y. Bai, 'Correlates and consequences of spanking and verbal punishment for low-income white, African-American, and Mexican-American toddlers', Child Development, 80: 5 (2009), 1403–20.

3 M. T. Wang and S. Kenny, 'Longitudinal links between fathers' and mothers' harsh verbal discipline and adolescents' conduct problems and depressive symptoms. Child development, 85:3 (2014), 908–23.

4 Dweck, C. S. Self-Theories: Their Role in Motivation, Personality and Development (Philadelphia: Psychology Press, 1999).

5 B. Estes and J. Wang, 'Workplace incivility: impacts on individual and organizational performance', Human Resource Development Review, 72 (2008) 218–240.

6 B. J. Tepper, 'Abusive supervision in work organizations: review, synthesis, and research agenda', Journal of Management, 33: 3 (2007), 261–89.

7 J. M. Gottman, 'Gottman method couple therapy', Clinical Handbook of Couple Therapy, 4: 8 (2008), 138–64.

## Chapter 9

1 https://mosaicscience.com/story/ surviving-troubled-childhood-resilience-neglect-adversity

2 Sequalae are conditions which are the consequence of a previous disease or injury.

3 E. E. Werner and R S. Smith, Overcoming the Odds: High-Risk Children from Birth to Adulthood (Ithaca, NY: Cornell University Press, 1992).

[4] E. E. Werner, 'What can we learn about resilience from large-scale longitudinal studies?' in S. Goldstein and R. Brooks (eds), *Handbook of Resilience in Children* (New York: Kluwer Academic Publishers, 2005), 91–106.

[5] Emmy Werner, 'Risk, resilience and recovery', http://faculty.mwsu.edu/psychology/dave.carlston/Child/Undergrad/resilience.pdf

[6] E. J. McCrory, M. I. Gerin and E Viding, 'Annual Research Review: childhood maltreatment, latent vulnerability and the shift to preventative psychiatry – the contribution of functional brain imaging', *Journal of Child Psychology and Psychiatry* (2017) 58(4): 338–357.

[7] P. Fonagy and E. Allison, 'The role of mentalising and epistemic trust in the therapeutic relationship', *Psychotherapy*, 51: 3 (2014), 372.

## Chapter 10

[1] S. H. Wang and R. Baillargeon, 'Inducing infants to detect a physical violation in a single trial', *Psychological Science*, 16: 7 (2005), 542–9.

[2] M. Donaldson and J. McGarrigle, 'Some clues to the nature of semantic development', *Journal of Child Language*, 1: 2 (1974), 185–94.

[3] A. Karmiloff-Smith, 'Nativism versus neuroconstructivism: rethinking the study of developmental disorders', *Developmental Psychology*, 45: 1 (2009), 56.

[4] Karmiloff-Smith, 'Nativism . . .', op. cit., 56.

[5] Literature reviewed by K. A. Ericsson, 'The influence of experience and deliberate practice on the development of superior expert performance', *Cambridge Handbook of Expertise and Expert Performance*, 38 (2006), 685–705.

## Chapter 11

[1] Law Lords Department, 'Judgments – Reg. v. Secretary of State for the Home Department, Ex parte V. and Reg. v. Secretary of State for the Home Department, Ex parte T', Publications.parliament.uk. Retrieved 15 January 2012.

[2] D. K. Lapsley, *Moral Psychology* (Westview Press, 1996).

[3] A. Blasi, 'Bridging moral cognition and moral action: a critical review of the literature', *Psychological Bulletin*, 88: 1 (1980), 1.

[4] W. Damon, *The Social World of the Child* (San Francisco: Jossey-Bass, 1977).

[5] J. K. Hamlin, K. Wynn and P. Bloom, 'Social evaluation by preverbal infants', *Nature*, 450: 7,169 (2007), 557–9.

[6] J. K. Hamlin and K. Wynn, 'Young infants prefer prosocial to antisocial others', *Cognitive Development*, 26: 1 (2011), 30–9.

[7] F. Warneken and M. Tomasello, 'Altruistic helping in human infants and young chimpanzees', *Science*, 311: 5,765 (2006), 1301–3.

[8] E. Turiel, 'The development of morality', *Handbook of Child Psychology* (New York: Wiley, 1998) Vol. 3, 5th edition, 863–932.

[9] Damasio et al., 'Moral emotion experienced following a transgression, or merely anticipated, is the motivational engine that infuses misdeeds with negative personal valence' (1994).

[10] J. Haidt, 'The moral emotions', *Handbook of Affective Sciences*, 11 (2003), 852–70.

[11] P. A. Miller and M. A. J. O. De Haar, 'Emotional, cognitive, behavioral, and temperament characteristics of high-empathy children', *Motivation and Emotion*, 21: 1 (1997), 109–25.

[12] P. A. Miller, N. Eisenberg, R. A. Fabes and R. Shell, 'Relations of moral reasoning and vicarious emotion to young children's prosocial behavior toward peers and adults', *Developmental Psychology*, 32: 2 (1996), 210.

[13] L. C. Findlay, A. Girardi and R. J. Coplan, 'Links between empathy, social behavior, and social understanding in early childhood', *Early Childhood Research Quarterly*, 21: 3 (2006), 347–59.

[14] D. Jolliffe and D. P. Farrington, 'Empathy and offending: a systematic review and meta-analysis', *Aggression and Violent Behavior*, 9: 5 (2004), 441–76.

[15] J. Stuewig, J. P. Tangney, C. Heigel, L. Harty and L. McCloskey, 'Shaming, blaming, and maiming: functional links among the moral emotions, externalization of blame, and aggression', *Journal of Research in Personality*, 44: 1 (2010), 91–102, and ideas discussed in lecture on moral development by Dr A. Searacardoso at Anna Freud Centre, 2014.

[16] J. Joireman, 'Empathy and the self-absorption paradox II: self-rumination and self-reflection as mediators between shame, guilt, and empathy', *Self and Identity*, 3: 3 (2004), 225–38.

[17] J. P. Tangney, J. Stuewig and D. J. Mashek, 'Moral emotions and moral behavior', *Annual Review of Psychology*, 58 (2007), 345–72.

[18] Interview with Essi Viding.

[19] Interview with Essi Viding.

## Chapter 12

[1] A. M. Minnett, D. L. Vandell and J. W. Santrock, 'The effects of sibling status on sibling interaction: influence of birth order, age spacing, sex of child, and sex of sibling', *Child Development* (1983), 1064–72.

[2] Minnett et al., 'The effects . . .', op. cit., 1064–72.

[3] C. Stocker, J. Dunn and R. Plomin, 'Sibling relationships: links with child temperament, maternal behavior, and family structure', *Child Development* (1989), 715–27.

[4] S. M. McHale, K. A. Updegraff and S. D. Whiteman, 'Sibling relationships and influences in childhood and adolescence', *Journal of Marriage and the Family*, 74: 5 (2012), 913–30.

[5] A. Piontelli, *From Fetus to Child: An Observational and Psychoanalytic Study* (London: Routledge, 2003).

[6] K. Thorpe and S. Danby, 'Compromised or competent: analysing twin children's social worlds', *Twin Research and Human Genetics*, 9: 1 (2006), 90.

[7] I. Deary, A. Pattie, V. Wilson and L. Whalley, 'The cognitive cost of being a twin: two whole-population surveys', *Twin Research and Human Genetics*, 8: 4 (2005), 376–83.

[8] P. R. Amato and B. Keith, 'Parental divorce and the well-being of children: a meta-analysis', *Psychological Bulletin*, 110: 1 (1991), 26–46.

[9] L. A. Gennetian, 'One or two parents? Half or step siblings? The effect of family structure on young children's achievement', *Journal of Population Economics*, 18: 3 (2005),

415–36.

## Chapter 13

[1] Several of the ideas in this chapter are influenced by a lecture on play given at the Anna Freud Centre in the years 2010–15 by Angela Joyce, child and adult psychoanalyst.

[2] W. W. Hartup, *Having Friends, Making Friends, and Keeping Friends: Relationships as Educational Contexts* (ERIC Digest, 1992).

[3] R. Tagore, 'On the Seashore, the Crescent Moon' (1957). Quotation cited by Angela Joyce in her lecture on play at the Anna Freud Centre.

[4] D. W. Winnicott, 'Playing: a theoretical statement' in *Playing and Reality*, (London: Routledge, 1971).

[5] J. Panksepp, 'Can play diminish ADHD and facilitate the construction of the social brain?' *Journal of the Canadian Academy of Child and Adolescent Psychiatry*, 16: 2 (2007), 57–66.

[6] https://www.theguardian.com/artanddesign/2014/aug/31/playing-to-the-gallery-grayson-perry-review

## Chapter 14

[1] S. H. Landry, K. E. Smith, P. R. Swank and C. L. Miller-Loncar, 'Early maternal and child influences on children's later independent cognitive and social functioning', *Child Development*, 71: 2 (2000), 358–75.

[2] R. C. Pianta, *Enhancing Relationships between Children and Teachers* (American Psychological Association, 1999).

[3] S. H. Birch and G. W. Ladd, 'Children's interpersonal behaviors and the teacher–child relationship', *Developmental Psychology*, 34: 5 (1998), 934.

[4] Birch and Ladd, 'Children's interpersonal behaviors . . .', op. cit., 934.

[5] M. Pianta, 'Technology and growth in OECD countries, 1970–90', *Cambridge Journal of Economics*, 19 (1995), 175–87.

[6] E. O'Connor and K. McCartney, 'Testing associations between young children's relationships with mothers and teachers', *Journal of Educational Psychology*, 98: 1 (2006), 87.

[7] E. W. Saft and R. C. Pianta, 'Teachers' perceptions of their relationships with students: effects of child age, gender, and ethnicity of teachers and children', *School Psychology Quarterly*, 16: 2 (2001), 125.

[8] C. C. Raver, S. M. Jones, C. P. Li-Grining, M. Metzger, K. M. Champion and L. Sardin, 'improving preschool classroom processes: preliminary findings from a randomized trial implemented in Head Start settings', *Early Childhood Research Quarterly*, 23: 1 (2008), 10-26.

[9] Bowlby, *Attachment: Attachment and Loss*, vol. 1, op. cit.

[10] O'Connor and McCartney, op. cit.

[11] C. Howes, 'Social-emotional classroom climate in child care, child–teacher relationships and children's second grade peer relations' *Social Development*, 9: 2 (2000), 191–204.

[12] A. T. Henderson and K. L. Mapp, 'A new wave of evidence: the impact of school, family, and community connections on student achievement. Annual synthesis 2002', (National

# Notes

Center for Family and Community Connections with Schools, 2002).

[13] W. T. Miedel and A. J. Reynolds, 'Parent involvement in early intervention for disadvantaged children: does it matter?' *Journal of School Psychology*, 37: 4 (2000), 379–402.

[14] J. S. Eccles and R. D. Harold, 'Parent–school involvement during the early adolescent years', *Teachers' College Record*, 94, (1993) 568–87.

[15] G. Crozier and J. Davies, 'Hard-to-reach parents or hard-to-reach schools? A discussion of home–school relations, with particular reference to Bangladeshi and Pakistani parents', *British Educational Research Journal*, 33: 3 (2007), 295–313.

[16] K. H. Corriveau, P. L. Harris, E. Meins, C. Fernyhough, B. Arnott, L. Elliott, B. Liddle, A. Hearn, L. Vitorrini and M. De Rosnay, 'Young children's trust in their mother's claims: longitudinal links with attachment security in infancy', *Child Development*, 80: 3 (2009), 750–61.

[17] K. Egyed, I. Király and G. Gergely, 'Communicating shared knowledge in infancy', *Psychological Science*, 24: 7 (2013), 1348–53.

[18] K. R. Wentzel and K. Caldwell, 'Friendships, peer acceptance, and group membership: reactions to academic achievement in middle school', *Child Development*, 68: 6 (1997), 1198–209.

[19] R. A. Fabes, L. D. Hanish and C. L. Martin, 'Children at play: the role of peers in understanding the effects of child care', *Child Development*, 74: 4 (2003), 1039–43.

[20] V. E. Besag, *Bullies and Victims in Schools: A Guide to Understanding and Management* (Buckingham: Open University Press, 1989).

[21] R. Kaltiala-Heino, M. Rimpelä, M. Marttunen, A. Rimpelä and P. Rantanen, 'Bullying, depression, and suicidal ideation in Finnish adolescents: school survey', *BMJ*, 319: 7,206 (1999), 348–51.

[22] http://www.pacer.org/bullying/resources/stats.asp

[23] K. Williams, M. Chambers, S. Logan and D. Robinson, 'Association of common health symptoms with bullying in primary school children', *BMJ*, 313: 7,048 (1996), 17–19.

[24] M. J. Boulton and K. Underwood, 'Bully/victim problems among middle-school children', *British Journal of Educational Psychology*, 62: 1 (1992), 73–87.

[25] R. Kaltiala-Heino et al., op. cit., 348–51.

[26] R. M. Kowalski, G. W. Giumetti, A. N. Schroeder and M. R. Lattanner, 'Bullying in the digital age: a critical review and meta-analysis of cyberbullying research among youth', *Psychological Bulletin*, 140: 4 (2014), 1073–137.

[27] L. R. Huesmann, L. D. Eron, M. M. Lefkowitz and L. O. Walder, 'Stability of aggression over time and generations', *Developmental Psychology*, 20: 6 (1984), 1120.

[28] L. D. Eron, L. R. Huesmann, E. Dubow, R. Romanoff, and P. W. Yarmel, 'Aggression and its correlates over 22 years', in *Childhood Aggression and Violence* (1987), 249–62.

[29] D. Olweus, Bullying at school: basic facts and effects of a school-based intervention program', *Journal of Child Psychology and Psychiatry*, 35: 7 (1994), 1171–90.

[30] P. T. Slee, 'Peer victimization and its relationship to depression among Australian primary school students', *Personality and Individual Differences*, 18: 1 (1995), 57–62.

[31] R. Kaltiala-Heino et al., op. cit., 348–51.

[32] *Why Be Happy When You Could Be Normal?* (London: Jonathan Cape, 2011), 54–5.

[33] A. M. Popp, A. A. Peguero, K. R. Day and L. L. Kahle, 'Gender, bullying victimization, and education', *Violence and Victims*, 29: 5 (2014), 843–56.

[34] J. J. Ratey, E. Hagerman and J. Ratey, *Spark: How Exercise Will Improve the Performance of Your Brain* (London: Quercus, 2010).

[35] https://www.youtube.com/watch?v=ERbvKrH-GC4

## Chapter 15

[1] L. Oakes, 'Using habituation of looking time to assess mental processes in infancy', *Journal of Cognition and Development*, 11: 3 (2010), 255–68.

[2] S. A. Rose, J. F. Feldman, J. J. Jankowski and R. Van Rossem, 'The structure of memory in infants and toddlers: an SEM study with full-terms and pre-terms', *Developmental Science*, 14: 1 (2011), 83–91.

[3] E. McCrory, lecture on memory at Anna Freud Centre, London, UK, 2014.

[4] M. A. Conway and C. W. Pleydell-Pearce, 'The construction of autobiographical memories in the self-memory system', *Psychological Review*, 107: 2 (2000), 261.

[5] J. Cassidy and R. P. Shaver, 'Handbook of attachment', *Theory, Research, and Clinical Applications*, (Guildford Press, 1999).

[6] E. McCrory lecture, op. cit.

[7] F. Jack, S. MacDonald, E. Reese and H. Hayne, 'Maternal reminiscing style during early childhood predicts the age of adolescents' earliest memories', *Child Development*, 80: 2 (2009), 496–505.

[8] K. Salmon and E. Reese, 'The benefits of reminiscing with young children', *Current Directions in Psychological Science*, 25: 4 (2016), 233–8.

[9] J. A. Sumner, J. W. Griffith and S. Mineka, 'Over-general autobiographical memory as a predictor of the course of depression: a meta-analysis', *Behaviour Research and Therapy*, 48: 7 (2010), 614–25.

[10] C. R. Brewin, 'Autobiographical memory for trauma: update on four controversies', *Memory*, 15: 3 (2007), 227–48.

[11] J. M. G. Williams, T. Barnhofer, C. Crane, Herman, F. Raes, E. Watkins and T. Dalgleish, 'Autobiographical memory specificity and emotional disorder', *Psychological Bulletin*, 133: 1 (2007), 122.

[12] G. S., Goodman and A. Melinder, 'Child witness research and forensic interviews of young children: a review', *Legal and Criminological Psychology*, 12: 1 (2007), 1–19.

## Chapter 16

[1] B. J. Ellis, S. McFadyen-Ketchum, K. A. Dodge, G. S. Pettit and J. E. Bates, 'Quality of early family relationships and individual differences in the timing of pubertal maturation in girls: a longitudinal test of an evolutionary model' *Journal of Personality and Social Psychology*, 77: 2 (1999), 387.

[2] Increases in sex hormone secretions can occur between six and fifteen, changes in bodily appearance can happen between eight and eighteen, and brain changes occur throughout childhood, teenage years and beyond.

# Notes

[3] P. Blos, 'The second individuation process of adolescence', *Psychoanalytic Study of the Child*, 22 (1967), 162–86.

[4] S. Briggs, *Working with Adolescents: A Contemporary Psychodynamic Approach* (London: Palgrave, 2002).

[5] H. R. Schaffer, 'Child psychology: the future', S. Chess and A. Thomas (eds), *Annual Progress in Child Psychiatry and Child Development* (NY: Brunner/Mazel, 1988).

[6] E.g. P. A. Sutherland, *Cognitive Development Today: Piaget and His Critics* (London: SAGE, 1992).

[7] J. N. Giedd, J. W. Snell, N. Lange, J. C. Rajapakse, B. J. Casey, P. L. Kozuch, A.C. Vaituzis, Y. C. Vauss, S. D. Hamburger, D. Kaysun and J. L. Rapoport, 'Quantitative magnetic resonance imaging of human brain development: ages 4–18', *Cerebral Cortex*, 6: 4 (1996), 551–9.

[8] D. Elkind, 'Egocentrism in adolescence', *Child Development*, 38 (1967), 1025–34.

[9] http://bayanbox.ir/view/3495143401881764256/rahpu.ir-TNC181.pdf

[10] B. F. Skinner, 'Shaping and maintaining operant behaviour': Chapter 6 of *Science and Human Behavior* (New York: Simon & Schuster, 1953), 91.

[11] A. Bandura, 'Human agency in social cognitive theory', *American Psychologist*, 44: 9 (1989), 1175.

[12] L. Steinberg, S. Graham, L. O'Brien, J. Woolard, E. Cauffman and M. Banich, 'Age differences in future orientation and delay discounting', *Child Development*, 80: 1 (2009), 28–44.

[13] Gardner and Steinberg, 2005, described in review: http://www.cla.temple.edu/tunl/publications/documents/Albert_Chein_CDPS_2013.pdf

[14] Ideas from P. Graham, *The End of Adolescence* (Oxford University Press, 2004).

## Chapter 17

[1] http://abcnews.go.com/Health/genderless-baby-controversy-mom-defends-choice-reveal-sex/story?id=13718047

[2] J. Cohen, 'A power primer', *Psychological Bulletin*, 112: 1 (1992), 155.

[3] J. Money and A. A. Ehrhardt, 'Man and woman, boy and girl: differentiation and dimorphism of gender identity from conception to maturity' (Oxford, England: Johns Hopkins University Press, 1972).

[4] M. L. Collaer and M. Hines, 'Human behavioral sex differences: a role for gonadal hormones during early development?' *Psychological Bulletin*, 118: 1 (1995), 55.

[5] Interview over email with Christia Spears Brown.

[6] Reviewed in C. L., Martin and D. Ruble, 'Children's search for gender cues: cognitive perspectives on gender development', *Current Directions in Psychological Science*, 13: 2 (2004), 67–70.

[7] A. J. Rose and K. D. Rudolph, 'A review of sex differences in peer relationship processes: potential trade-offs for the emotional and behavioral development of girls and boys', *Psychological Bulletin*, 132: 1 (2006), 98.

[8] E. E. Maccoby and C. N. Jacklin, 'Gender segregation in childhood', *Advances in Child Development and Behavior*, 20 (1987), 239–87.

[9] N. M. Else-Quest, J. S. Hyde, H. H. Goldsmith and C. A. Van Hulle, 'Gender differences in temperament: a meta-analysis', *Psychological Bulletin*, 132: 1 (2006), 33.

[10] J. Archer, 'Sex differences in aggression in real-world settings: a meta-analytic review', *Review of General Psychology*, 8: 4 (2004), 291.

[11] W. O. Eaton and L. R. Enns, 'Sex differences in human motor activity level', *Psychological Bulletin*, 100: 1 (1986), 19.

[12] R. S. Bigler, A. E. Arthur, J. M. Hughes and M. M. Patterson, 'The politics of race and gender: children's perceptions of discrimination and the US presidency', *Analyses of Social Issues and Public Policy*, 8 (2008), 83–112.

[13] Rose and Rudolph, op. cit., 98.

[14] F. Poulin and S. Pedersen, 'Developmental changes in gender composition of friendship networks in adolescent girls and boys', *Developmental Psychology*, 43: 6 (2007), 1484.

[15] C. Leaper, M. Carson, C. Baker, H. Holliday and S. Myers, 'Self-disclosure and listener verbal support in same-gender and cross-gender friends' conversations', *Sex Roles*, 33: 5–6 (1995), 387–404.

[16] M. A. Messner, 'Barbie girls versus sea monsters: children constructing gender', *Gender and Society*, 14: 6 (2000), 765–84.

[17] N. A. Card, B. D. Stucky, G. M. Sawalani and T. D. Little, 'Direct and indirect aggression during childhood and adolescence: a meta-analytic review of gender differences, intercorrelations, and relations to maladjustment', *Child Development*, 79: 5 (2008), 1185–229.

[18] J. E. O. Blakemore, S. A. Berenbaum and L. S. Liben, *Gender Development* (Chicago: Psychology Press, 2008).

[19] B. K. Todd, J.A. Barry and S. A. O. Thommessen, 'Preferences for "gender-typed" toys in boys and girls aged 9 to 32 months', *Infant and Child Development*, 26: 3, e1986 (2017).

[20] G. M. Alexander and M. Hines, 'Sex differences in response to children's toys in non-human primates (*Cercopithecus aethiops sabaeus*)', *Evolution and Human Behavior*, 23: 6 (2002), 467–79.

[21] J. M. Hassett, E. R. Siebert and K. Wallen, 'Sex differences in rhesus monkey toy preferences parallel those of children', *Hormones and Behavior*, 54: 3 (2008), 359–64.

[22] K. C. Kling, J. S. Hyde, C. J. Showers and B. N. Buswell, 'Gender differences in self-esteem: a meta-analysis', *Psychological Bulletin*, 125: 4 (1999), 470.

[23] J. M. Twenge and S. Nolen-Hoeksema, 'Age, gender, race, socioeconomic status, and birth cohort difference on the children's depression inventory: a meta-analysis', *Journal of Abnormal Psychology*, 111: 4 (2002), 578–88.

[24] S. Grabe, L. M. Ward and J. S. Hyde, 'The role of the media in body image concerns among women: a meta-analysis of experimental and correlational studies', *Psychological Bulletin*, 134: 3 (2008), 460.

[25] A. Feingold and R. Mazzella, 'Gender differences in body image are increasing', *Psychological Science*, 9: 3 (1998), 190–5.

[26] D. C. Jones and J. K. Crawford, 'The peer appearance culture during adolescence: gender and body mass variations', *Journal of Youth and Adolescence*, 35: 2 (2006), 243.

[27] M. Ingalhalikar, A. Smith, D. Parker, T. D. Satterthwaite, M. A. Elliott, K. Ruparel, H. Hakonarson, R. E. Gur, R. C. Our and R. Verma, 'Sex differences in the structural connectome of the human brain', *Proceedings of the National Academy of Sciences*, 111: 2 (2014), 823–8.

[28] D. L. Best and J. J. Thomas, 'Cultural diversity and cross-cultural perspectives', in A. H. Eagly, A. E. Beall and R. J. Sternberg (eds), *The Psychology of Gender*, 2nd edn (New York: Guilford, 2004), 296–327.

[29] C. Leaper, K. J. Anderson and P. Sanders, 'Moderators of gender effects on parents' talk to their children: a meta-analysis', *Developmental Psychology*, 34:1 (1998), 3–27.

[30] H. Lytton and D. M. Romney, 'Parents' differential socialization of boys and girls: a meta-analysis', *Psychological Bulletin*, 109: 2 (1991), 267.

[31] R. M. Ryan and J. H. Lynch, 'Emotional autonomy versus detachment: revisiting the vicissitudes of adolescence and young adulthood', *Child Development*, 60: 2 (1989), 340–56.

[32] R. Chao and V. Tseng, 'Parenting of Asians', *Handbook of Parenting*, 4 (2002), 59–93.

[33] C. Leaper and T. E. Smith, 'A meta-analytic review of gender variations in children's language use: talkativeness, affiliative speech, and assertive speech', *Developmental Psychology*, 40: 6 (2004), 993.

[34] Jones and Crawford, op. cit, 243.

[35] M. A. Messner, 'The limits of "the male sex role": an analysis of the men's liberation and men's rights movements discourse', *Gender & Society* 12: 3 (1998), 255–76.

[36] N. A. Card, B. D. Stucky, G. M. Sawalani and T. D. Little, 'Direct and indirect aggression during childhood and adolescence: a meta-analytic review of gender differences, intercorrelations, and relations to maladjustment', *Child Development*, 79: 5 (2008), 1185–229.

[37] R. S. Bigler, 'The role of classification skill in moderating environmental influences on children's gender stereotyping: a study of the functional use of gender in the classroom', *Child Development*, 66: 4 (1995), 1072–87.

[38] A. Levy, *Female Chauvinist Pigs* (London: Free Press, 2006).

[39] A. Bandura, 'The explanatory and predictive scope of self-efficacy theory', *Journal of Social and Clinical Psychology*, 4: 3 (1986), 359–73.

[40] K. Bussey and A. Bandura, 'Social cognitive theory of gender development and differentiation', *Psychological Review*, 106: 4 (1999), 676.

[41] Leaper and Smith, 'A meta-analytic review . . .', op. cit., 993.

[42] Leaper, Anderson and Sanders, op. cit.

[43] C. Leaper and M. M. Ayres, 'A meta-analytic review of gender variations in adults' language use: talkativeness, affiliative speech, and assertive speech', *Personality and Social Psychology Review*, 11: 4 (2007), 328-63.

## Chapter 18

[1] J. E. Marcia, 'Identity and psychosocial development in adulthood', *Identity: An International Journal of Theory and Research*, 2 (2002), 7–28.

[2] A. Keller, L. M. Ford and M. A. Meacham, 'Dimensions of self-concept in pre-school children', *Developmental Psychology*, 14 (1978), 483–9.

[3] R. A. Eder, 'Uncovering children's psychological selves: individual and developmental differences', *Child Development*, 61 (1990), 849–63.

[4] J. Kroger, 'Why is identity achievement so elusive?' *Identity: An International Journal of Theory and Research*, 7: 4 (2007), 331–48.

[5] An important caveat to these findings is that there are several different cortical midline structures, and studies vary to some degree in what they count as being the cortical midline structures or not, so it's worth taking any study that is too certain about this being the 'seat of identity' with a healthy pinch of salt. The list of what the cortical midline structures are supposed to do is also pretty long, including: representation, monitoring self-referential stimuli, evaluation, integration, self-awareness, agency, knowing where we are in physical space, understanding ownership, mind-reading of other people, emotion, and autobiographical memory. Fair enough that these things have all got some relation to our sense of self, but they also relate to a fair amount of other things.

[6] They found four clusters of brain activity that differed, including the CMS but also the insula, temporal pole and the inferior frontal cortex.

[7] The anterior rostral and dorsal medial prefrontal cortex and the dorsal, rostral and ventral ACC.

[8] These might include the temporo-parietal junction (TPJ) which is involved in third-person perspective-taking and theory-of-mind tasks (remember Sally–Anne from Chapter 7), and the temporal poles which are involved in memories relating to ourselves. For more see Jennifer H. Pfeifer and Shannon J. Peake, 'Self-development: integrating cognitive, socioemotional and neuroimaging perspectives', *Developmental Cognitive Neuroscience* (2011) 55–69.

## Chapter 19

[1] J. Burnham, 'Systemic supervision: the evolution of reflexivity in the context of the supervisory relationship', *Human Systems*, vol. 4 (1993), 349–81.

# Index

Page numbers in *italic* refer to Figures.

# Index

and attachment relationships 104–5
fallibility 101–2
mentalisation-based treatment 102–4,
107
milestones 43–6, 51, 257, 277
adult 46, 51
cultural differences 45
developmental differences 45
family transitions 47–8
language acquisition 82
play and 181
sibling relationships, effects of 174
Milgram study 2
mind-reading *see* mentalisation; theory of
mind
mindfulness-based therapies 283
mind–body connection 65, 67, 68
mirror test 3, 262
Mischel, Walter 148
Monk, Catherine 18
moral development 152–67, 180, 278
environmental influences 152
moral cognition 153–4
moral processing 160, *161*, 166
stage theory of 155, 156, 180
moral dilemmas 97, 154–5, 156–7, 180
moral emotions 157–8, *159*, 166
moral sense 152, 154, 162
absence of 159–60
emotions, role of 157
fostering 164–7
male and female perspectives 155
Music, Graham 147–8, 174–5, 184,
207–8

NanoWriMo 121
narrative therapy 284–5
nature–nurture debate 20, 59, 65, 67, 260,
277
naughty step 110–11, 114
near infrared spectroscopy (NIRS) 83
Nelson, Catherine 215
Network Relationship Inventory 268
neuroconstructivism 146–7
neurodevelopment 65
neurogenesis 5–6

neuronal pruning 239
neurotransmitters 20, 38, 252
*see also specific index entries*
noradrenaline 19, 20, 37
'normal', accepted ideas of 41, 46
novel objects task 199–202, *200*, *201*

object permanence 25–6, 140–1
obsessive compulsive disorder 37
occipital lobe 70, *70*, *161*
oestrogens 248, 258
Olweus, Dan 204
only children *see* singletons
open-ended questions 222
operant conditioning 109–10, 119
optic nerve 70
orbitofrontal cortex 133, 160, *161*
oxytocin 37–8

Panksepp, Jaak 185
parenting programmes 113–14
parenting pyramid *113*, 114
parenting styles 114–18, *115*
authoritarian 115–16
authoritative 117, 118
indulgent 116–17
low and high demandingness 114, 115,
116, 117
low and high responsiveness 114, 115,
116, 117
neglectful 117
Pavlov, Ivan 108–9
peer relationships 174, 195, 202–6
gender-normative behaviour 253
peer learning 180
peer pressure 235, 237
*see also* friendships
pendulum task 145
peripheral nervous system 8
Perry, Grayson 190
perspective-taking 155, 186, 280
*see also* theory of mind
phobias 109
Piaget, Jean 139–41, 142, 144–5, 150,
153–4, 156, 179–80, 230–2
play 179–93, 279

# Index